MAJOR EVENTS IN
THE HISTORY OF LIFE

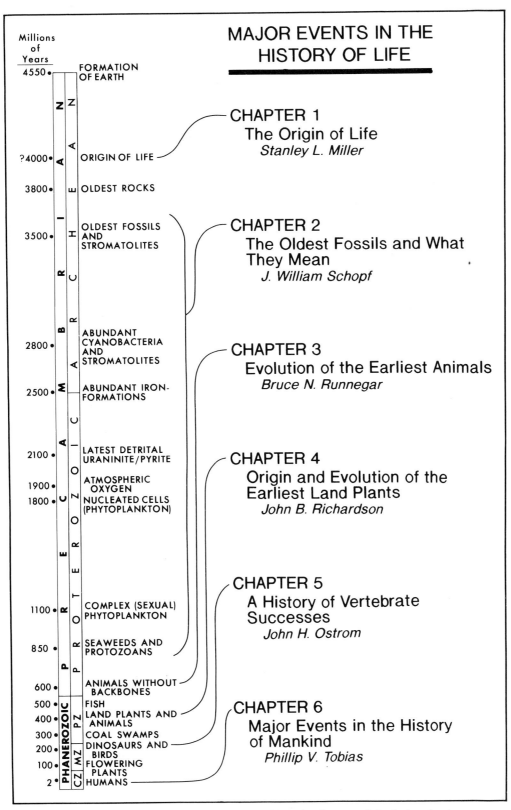

MAJOR EVENTS IN THE HISTORY OF LIFE

CHAPTER 1
The Origin of Life
Stanley L. Miller

CHAPTER 2
The Oldest Fossils and What
They Mean
J. William Schopf

CHAPTER 3
Evolution of the Earliest Animals
Bruce N. Runnegar

CHAPTER 4
Origin and Evolution of the
Earliest Land Plants
John B. Richardson

CHAPTER 5
A History of Vertebrate
Successes
John H. Ostrom

CHAPTER 6
Major Events in the History
of Mankind
Phillip V. Tobias

Geologic time-scale, in millions of years before the present, illustrating the temporal distribution of major events in the history of life.

MAJOR EVENTS IN THE HISTORY OF LIFE

EDITED BY

J. WILLIAM SCHOPF

JONES AND BARTLETT PUBLISHERS

Boston

Editorial, Sales, and Customer Service Offices
Jones and Bartlett Publishers
20 Park Plaza
Boston, MA 02116

Library of Congress Cataloging-in-Publication Data

Major events in the history of life / J. William Schopf, editor.
    p.   cm.
    Proceedings from a symposium convened by the IGPP Center for the Study of Evolution and the Origin of Life at the University of California, Los Angeles, held January 11, 1991.
    Includes bibliographical references and index.
    ISBN 0-86720-268-8
    1. Evolution (Biology)—Congresses. 2. Paleontology—Congresses.
I. Schopf, J. William, 1941- II. University of California, Los Angeles, Center for the Study of Evolution and the Origin of Life.
    QH359.M35   1992
    575—dc20

                                        91-36263
                                        CIP

ISBN 0-86720-268-8

Production Editor: Joni Hopkins McDonald
Editorial Production Service: York Production Services
Design: Image House, Inc./Stuart D. Paterson
Cover Design: Hannus Design Associates
Printing and Binding: Courier Westford

A contribution of the IGPP Center for the Study of Evolution and the Origin of Life (CSEOL), University of California, Los Angeles.

Printed in the United States of America
95 94 93 92 91   10 9 8 7 6 5 4 3 2 1

CONTENTS

CHAPTER 2   **THE OLDEST FOSSILS AND WHAT THEY MEAN   29**

J. William Schopf

PREFACE

More than 1700 students, faculty, and other members of the UCLA community attended a "Major Events in the History of Life" symposium on January 11, 1991, convened by the IGPP Center for the Study of Evolution and the Origin of Life at the University of California, Los Angeles. This volume makes accessible the proceedings of that symposium.

Presented at a level appropriate for use in first- or second-year college coursework, the six chapters of this book summarize our understanding of crucial events that shaped the development of the earth's environment and the course of biological evolution over some four billion years of geologic time. As illustrated in the frontispiece, the subjects covered by acknowledged leaders in their fields span an enormous sweep of biologic history, from the formation of planet Earth and the origin of living systems to our earliest records of human activity. Several chapters present new data and new syntheses, or summarize results of new types of analysis, material not usually available in current college textbooks, and some of them are broadly multidisciplinary, calling on data from diverse areas of science. Taken together, these chapters illustrate a cardinal truth: Unlike the traditional curricular structure of colleges and universities, nature is not compartmentalized into the familiar academic disciplines such as biology, geology, and chemistry. A multidisciplinary approach seems certain to yield far better understanding of the history of life than that provided by any single discipline alone.

Biographical sketches of the authors of the six chapters of the book follow.

BIOGRAPHICAL SKETCHES

Chapter 1: The Prebiotic Synthesis of Organic Compounds as a Step Toward the Origin of Life
STANLEY L. MILLER

A native of Oakland, California, Dr. Miller received his undergraduate training at the University of California, Berkeley, and his doctorate in chemistry, in 1954, from the University of Chicago. Following completion of a F. B. Jewett Fellowship at the California Institute of Technology, and after serving as a faculty member in the College of Physicians and Surgeons at Columbia University, he joined the faculty of the University of California, San Diego, in 1958, where he is Professor of Chemistry. A member of the U.S. National Academy of Sciences, and recipient both of the A. I. Oparin Medal of the International Society for the Study of the Origin of Life and of the Distinguished Scientist Award of the IGPP Center for the Study of Evolution and the Origin of Life, Professor Miller is renowned for his classic investigations of chemical reactions on the primordial planet, studies that provide the foundation for current understanding of the origin of life on earth.

Stanley L. Miller (University of California, San Diego).

Chapter 2: The Oldest Fossils and What They Mean
J. WILLIAM SCHOPF

J. William Schopf (University of California, Los Angeles).

Dr. Schopf was educated at Oberlin College, Ohio, where he majored in geology, and at Harvard University, where he received his advanced degrees in biology. In 1968, he joined the faculty of the University of California, Los Angeles, where he is Professor of Paleobiology in the Department of Earth and Space Sciences, a member of the Molecular Biology Institute and of the Institute of Geophysics and Planetary Physics, and Director of the IGPP Center for the Study of Evolution and the Origin of Life. A member of the American Philosophical Society and a Fellow of the American Academy of Arts and Sciences, he has received two Guggenheim Fellowships and medals awarded by the U.S. National Science Board, the U.S. National Academy of Sciences, and the International Society for the Study of the Origin of Life. Professor Schopf's pioneering research focuses on the oldest fossil evidence of life and on the interrelated evolution of the earth's biosphere and its environment during Precambrian time, the earliest seven-eighths of geologic history.

Chapter 3: Evolution of the Earliest Animals
BRUCE RUNNEGAR

Bruce Runnegar (University of California, Los Angeles).

An Australian citizen and American resident, Dr. Runnegar received his higher education degrees (B.Sc., 1962; B.Sc.Hons, 1964; Ph.D., 1967; D.Sc., 1978) at the University of Queensland, Australia. From 1968 through 1987 he was a member of the faculty of the Department of Geology at the University of New England (Armidale, Australia), where he was awarded a Personal Chair in Geology in 1985. In 1987, Dr. Runnegar assumed his current position at the University of California, Los Angeles, where he is Professor of Geology in the Department of Earth and Space Sciences, and a member both of the Molecular Biology Institute and of the Institute of Geophysics and Planetary Physics. A Fellow of the Australian Academy of Science, he is also recipient of the academy's Mawson Medal. Professor Runnegar is an internationally recognized leader in studies of the oldest known animal fossils and of the evolutionary relationships among early evolving animal phyla.

Chapter 4: Origin and Evolution of the Earliest Land Plants
JOHN B. RICHARDSON

John B. Richardson (The Natural History Museum, London).

A British citizen, Dr. Richardson received his advanced degrees (B.Sc., 1956; Ph.D., 1960; D.Sc., 1990) at the University of Sheffield, England. In 1959, he joined the faculty of the Department of Geology of King's College, University of London, an association he maintained until 1978 when he assumed his present position of Principal Scientific Officer at the British Museum (Natural History). Since 1984, he has also been a Research Professor in the Center for Evolution and the Paleoenvironment at the State University of New York, Binghamton. A USSR Academy of Sciences Senior Visiting Scientist and a Fellow of the Geological Society (London), Dr. Richardson is an acknowledged authority on the evolution, biogeography, and paleoenvironmental distribution of the earliest land plants as evidenced in the fossil record by their dispersed spores.

Chapter 5: A History of Vertebrate Successes
JOHN H. OSTROM

Born in New York City, Dr. Ostrom received his undergraduate training at Union College, Schenectady, New York, and his doctorate in geology, from Columbia University, in 1960. After serving on the staff of the American Museum of Natural History (1951–1956) and as a faculty member of Brooklyn College, New York (1956–1956), and of Beloit College, Wisconsin (1956–1961), Dr. Ostrom joined the Department of Geology and Geophysics at Yale University where he is Professor of Geology and Curator of Vertebrate Paleontology at the Peabody Museum of Natural History. Professor Ostrom is recipient of a Guggenheim Fellowship, the Alexander von Humboldt Senior Scientist Award (Germany), and the F. V. Hayden Geological Medal of the Philadelphia Academy of Science. A prolific author and exemplary teacher, Dr. Ostrom is internationally known for his studies of dinosaurs and other ancient reptiles, *Archaeopteryx* and the origin of birds, and of the early evolution of vertebrate animals.

John H. Ostrom (Yale University, New Haven, Connecticut).

Chapter 6: Major Events in the History of Mankind
PHILLIP V. TOBIAS

Born in Durban, South Africa, Dr. Tobias has spent virtually his entire academic career at the University of the Witwatersrand, first as a student (B.Sc., 1946; B.Sc.Hons, 1947; M.B.B.Ch., 1950; Ph.D., 1953; D.Sc. 1967) and since 1951 as a member of the faculty. Since 1959 he has been Head of the Department of Anatomy and Human Biology, and since 1965, Director of the Palaeo-anthropology Research Unit and its predecessors. A widely acclaimed scholar, Professor Tobias is recipient of the Balzan International Prize, of medals awarded by six learned societies, and of honorary degrees from seven universities. An Honorary Fellow of the Royal Society of South Africa and of the Royal Anthropological Society of Great Britain and Ireland, he is also an Honorary Foreign Associate of the U.S. National Academy of Sciences and of the American Academy of Arts and Sciences. Professor Tobias is author of nine books and more than 700 other publications; he has been a pioneering, prolific contributor to current understanding of the evolutionary history of the hominids.

Phillip V. Tobias (University of the Witwatersrand, Johannesburg).

THE PREBIOTIC SYNTHESIS OF ORGANIC COMPOUNDS AS A STEP TOWARD THE ORIGIN OF LIFE

■

Stanley L. Miller†

■

INTRODUCTION

It is now generally accepted that life arose on the earth early in its history. The sequence of events started with the synthesis of simple organic compounds by various processes. These simple organic compounds reacted to form polymers, which in turn reacted to form structures of greater and greater complexity until one was formed that could be called living. This is a relatively new idea, first expressed clearly by Oparin (1938) with contributions by Haldane (1929), Urey (1952), and Bernal (1951). It is sometimes referred to as either the Oparin-Haldane or the Heterotrophic hypothesis (Horowitz, 1945). Older ideas that are no longer regarded seriously include the seeding of the earth from another planet (panspermia), the origin of life at the present time from decaying organic material (spontaneous generation), and the origin of an organism early in the earth's history by an extremely improbable event. The latter process assumed that the primitive earth was essentially the same as at present, except possibly for the absence of molecular oxygen, and that the first organism had to have been **autotrophic.** That is, it had to have been photosynthetic, using light energy to synthesize all of its organic compounds from carbon dioxide (CO_2) and water (H_2O).

One of Oparin's major contributions was to propose that the first organisms were **heterotrophic,** that is, they utilized prebiotically produced organic compounds available in the environment. The organisms still had to build proteins, nucleic acids, and other complex organic compounds within their cells, but they did not have to synthesize de novo the amino acids, purines, pyrimidines, and sugars that make up such biopolymers. Oparin as well as Urey also proposed that the early earth had a reducing atmosphere composed of methane (CH_4), ammonia (NH_3), water vapor (H_2O), and molecular hydrogen (H_2), and that organic compounds might be synthesized in such an atmosphere. On the basis of Urey's and Oparin's ideas, it was shown that amino acids could be synthesized in surprisingly high yield from the action of spark, simulating lightning discharges, on just such a strongly

†Department of Chemistry, University of California, San Diego, La Jolla, California 92093-0317 USA

reducing atmosphere (Miller, 1953). There is now a large literature dealing with the prebiotic synthesis of organic compounds and polymers, which is too extensive to review here so we cover only the highlights (for reviews, see Kenyon and Steinman, 1969; Lemmon, 1970; Miller and Orgel, 1974).

We first consider the geological setting and the time available for the origin of life. Next is a discussion of the synthesis of simple organic compounds (monomers) on the primitive earth, as well as on asteroids and on dust grains in interstellar space, followed by a consideration of the polymerization processes needed to combine these monomers into polymers. The chapter finishes with a description of template polymerization and a review of outstanding and unresolved questions.

GEOLOGICAL CONSIDERATIONS

The Time Available for the Origin of Life

The time period in which prebiotic syntheses of organic compounds took place is frequently misunderstood. The earth is 4550 million years (Ma) old, and the earliest fossils known, the Warrawoona microfossils and stromatolites, are about 3500 Ma in age (see Chapter 2). The difference is more than one billion years, but the time available for life to arise was probably much shorter. It is usually said that it took at least a few hundred million years for organisms to evolve to the level of those found in the Warrawoona sediments, but this evolution may have been much faster. Less time would have been available for the origin of life if the earth melted completely early in its history, destroying all earlier synthesized prebiotic organic compounds, or if the early earth was heated as a result of impacts of large asteroids, possibilities that we discuss here.

Melting of the Primitive Earth

There are many theories that seek to describe the events which took place during the condensation of cosmic dust and larger objects to form the earth. At one extreme, it is believed that the accumulation of material took place in less than 100,000 years; in this scheme, gravitational energy was also released sufficiently rapidly to melt the entire earth, including material at its surface. At the other extreme, the accretion of material is believed to have been sufficiently slow so that the gravitational energy released during accumulation was dissipated by radiation at a rate comparable to the rate of energy production. In this model, the interior of the earth would have melted due to the effects of adiabatic compression and the decay of radioactive elements, but the crust would not have melted.

Whether the entire earth was ever molten or not does not greatly affect our discussion of the prebiotic synthesis of organic compounds. If the entire earth did melt, then all organic compounds, both in the interior and on the surface, would have been severely heated and pyrolyzed completely to an equilibrium mixture consisting largely of CO_2 (carbon dioxide), CO (carbon monoxide), N_2 (molecular nitrogen), CH_4, H_2, NH_3, and H_2O. All organic compounds synthesized in the solar nebula and reaching the surface of the earth intact would have been destroyed on a molten earth. When the earth had cooled sufficiently, a crust would have formed. When the average temperature on the surface and in the atmosphere became low enough, organic compounds would have been synthesized and, most importantly, would have accumulated. It is possible that some of the organic compounds synthesized in the solar nebula were brought to the earth with dust particles, meteorites, and comets, and that they survived their impact with the atmosphere and with the earth's surface. It is not clear how much of a contribution these compounds made to the inventory of prebiotic organic compounds. Relatively unstable compounds, such as sugars and certain amino acids, cannot be accounted for in this manner, since they would have decomposed before life could arise. There would have been a continuous addition of organic matter brought to the earth by carbonaceous chondrites (a class of meteorites containing appreciable amounts of organic matter) after the initial major accretion, but their addition would not result in the accumulation of unstable organic compounds. In any case, unstable organic compounds, such as sugars, have not been found in carbonaceous chondrites. For unstable compounds to accumulate, a continuous synthesis is required; this means that such compounds must have been synthesized in the atmosphere or oceans of the primitive earth.

Impact Frustration of the Origin of Life

It has been realized recently that because the moon was bombarded by large objects until about 3800 Ma ago, the earth must have been similarly bombarded. A 10-km asteroid, which is thought to have hit the earth 65 Ma ago, may have played a role in the extinction of the dinosaurs and many other species. But during this event, the heating of the earth as a whole would have been small. However, larger asteroids, 50 to 60 km in diameter, are much more destructive. A 60-km object would supply enough energy to boil the ocean and kill all but the hardiest thermophilic bacteria (Maher and Stevenson, 1988). Higher temperatures would be produced by larger objects impacting the early earth, temperatures high enough to destroy even prebiotic organic compounds dissolved in the ocean. Thus, any living systems that may have originated very early in earth history could not have persisted until the end of the major bombardment, and prebiotic organic compounds also could not have accumulated until then. This only presents problems if it is thought that the origin of life took a long period, for example, on the order of one billion years.

A period of, for example, 500 Ma in duration does not, in my opinion, present any problems. Many writers have stressed that several improbable events were required for the origin of life, and that therefore much time was needed. It should be kept in mind that periods on the order of a billion years are so far removed from our experience that we have no feeling or judgment as to what is likely or unlikely to take place, especially with regard to organic chemical reactions. There are a number of reasons to think that life must have arisen in ten million years or less, based on the known rate of decomposition of organic compounds. Since we do not know the details of the processes required for the origin of life, we cannot say how much less time may have been involved, but a period of perhaps 10,000 years for such processes is not impossible.

Temperature of the Primitive Earth

The necessity to accumulate organic compounds for the synthesis of the first organism requires a low-temperature earth; otherwise, the organic compounds would decompose. The **half-lives** (the amount of time required for one-half of a given compound to undergo a particular chemical process, a standard way of indicating whether the process occurs slowly or rapidly) for decomposition at 25°C vary from several billion years for the amino acid alanine, to a few million years for the amino acid serine, to thousands of years for hydrolytic decomposition of peptides and polynucleotides, to a few hundred years for sugars. Because the temperature coefficients of these decompositions are large, the half-lives would be much less at temperatures of 50° or 100°C. Conversely, the half-lives would be longer at 0°C. The rates of hydrolysis of peptide and polynucleotide polymers, and of decomposition of sugars, are so large that it seems impossible that such compounds could have accumulated in aqueous solution and been used in the first organism, unless the temperature was low.

The temperature of the present ocean averages 4°C, with surface waters being somewhat warmer. The freezing point of distilled water is 0°C, whereas, because of its salt content, seawater begins to freeze at -1.8°C and solidifies almost completely at -21°C. The temperature of the primitive ocean is not known, but it can be said that the instability of various organic compounds and polymers suggests strongly that life could not have arisen in the ocean unless the temperature was below 25°C. A temperature of 0°C would have helped greatly, and -21°C would have been even better. At such low temperatures, most of the water on the primitive earth would have been in the form of ice, with liquid seawater confined to the equatorial oceans. These considerations would not apply if life arose very rapidly (within a period of 10,000 to 100,000 years), although some compounds (e.g., sugars) would not survive for more than a few hundred years.

There is another reason for believing that life originated at low temperatures, whether in the oceans or in lakes. All of the template-directed reactions (discussed later) that must have led to the emergence of biological organization take place only below the melting temperature of the appropriately organized (double helix) polynucleotide structure. These temperatures range from 0°C, or lower, to a maximum of perhaps 35°C in the case of polynucleotide-mononucleotide helices.

The environment in which life arose is frequently referred to as a warm dilute soup of organic compounds. The preceding discussion indicates that a cold concentrated soup would have provided a better environment for the origin of life. At low temperatures, the decomposition of organic compounds and polymers is slowed greatly. Although, at first sight, low temperatures might seem to have been disadvantageous (because chemical syntheses would have proceeded more slowly), in fact they may have been advantageous. It is the *ratios* of the rates of synthesis to the rates of decomposition that are important, rather than the absolute rates, if ample time is available. Since the temperature coefficients of the synthetic reactions are generally less than those of the decomposition reactions, low temperatures would have favored the synthesis of more complex organic compounds and polymers.

Oxygen in the Primitive Atmosphere

Molecular oxygen (O_2) is usually assumed to have been absent from the primitive atmosphere. One reason is that the major source of free oxygen, prior to the advent of O_2-producing photosynthetic organisms, would have been the photodissociation of water vapor in the upper atmosphere. This is not a large source of O_2, and the oxygen formed would have reacted with Fe^{2+} and other unoxidized inorganic reactants available in the environment (see Chapter 2). A second reason is that O_2 reacts relatively rapidly with organic compounds, especially in the presence of ultraviolet light, and would therefore have been consumed by reaction with prebiotic organic matter. These and related arguments are so compelling that it does not seem possible that organic compounds remained in the primitive ocean for any length of time after large amounts of O_2 entered the earth's atmosphere. Organic compounds are present on the surface of the earth only because they are continuously being resynthesized by living organisms. Organic compounds occur below the surface of the earth, for example, in coal and petroleum deposits, because these environments are anaerobic. It appears certain, therefore, that substantial amounts of O_2 were absent from the earth's atmosphere during the period when organic compounds were synthesized up to the time when the first organism evolved.

The Composition of the Primitive Atmosphere

There is only limited agreement regarding the probable constituents of the primitive atmosphere. Because no rocks older than 3800 Ma in age are known, there is no geological evidence concerning the conditions on the earth from the time of its formation, 4550 Ma ago, to the time of deposition of these oldest known rocks. Even the 3800-Ma-old Isua rocks in Greenland are not sufficiently well preserved to provide much evidence about the atmosphere at that time. Proposed atmospheres and the reasons given to favor them are not discussed here. However, as we show later, the more reducing atmospheres are more favorable for the synthesis of organic compounds, both in terms of the amount of yield and of the variety of compounds obtained. Such considerations cannot prove that the earth had a particular type of primitive atmosphere. There is considerable opinion that strongly reducing conditions were never present on the primitive earth, but this would mean that the organic compounds must have been brought in on comets and meteorites, and this assumption has its own set of problems.

ENERGY SOURCES

A wide variety of energy sources has been utilized with various gas mixtures since the first experiments using electric discharges. The importance of a given energy source is determined by the product of the energy available and its efficiency for organic compound synthesis. Since neither factor can be evaluated with precision, only a qualitative assessment of the energy sources can be made. We must emphasize that neither any single source of energy nor any single process is likely to account for all the organic compounds on the primitive earth (Miller et al., 1976). An estimate of the sources of energy on the earth at the present time is given in Table 1.1

TABLE 1.1 **Present Sources of Energy Averaged over the Earth.**

	Energy	
Source	**(cal/cm^2/yr)**	**(J/cm^2/yr)**
Total radiation from sun	260,000	1,090,000
Ultraviolet light		
<3000 Å	3,400	14,000
<2500 Å	563	2,360
<2000 Å	41	170
<1500 Å	1.7	7
Electric discharges	4[a]	17
Cosmic rays	0.0015	0.006
Radioactivity (to 1.0 km depth)	0.8	3.0
Volcanoes	0.13	0.5
Shock waves	1.1[b]	4.6

[a] 3 cal/cm^2/yr of corona discharge + 1 cal/cm^2/yr of lightning.

[b] 1 cal/cm^2/yr of this is in the shock wave of lightning bolts and is also included under electric discharges.

Ultraviolet light was probably the largest source of energy on the primitive earth. The wavelengths absorbed by the constituents of the primitive atmosphere are all below 2000 Å except for ammonia (<2300 Å) and hydrogen sulfide, H_2S (<2600 Å). Whether ultraviolet light was the most effective energy source for production of organic compounds is not clear. Most of the ultraviolet-induced photochemical reactions would occur in the upper atmosphere, and the products formed would, for the most part, absorb the longer wavelengths and therefore be decomposed before they reached the oceans, where they would be dissolved and protected from photochemical destruction. The yield of amino acids from the photolysis of CH_4, NH_3, and H_2O at wavelenghts between 1470 and 1294 Å is quite low (Groth and von Weyssenhoff, 1960), probably due to the low yields of hydrogen cyanide, HCN (an intermediate compound in the production of amino acids). The synthesis of amino acids by the photolysis of CH_4, C_2H_6 (ethane), NH_3, H_2O, and H_2S mixtures by ultraviolet light of wavelengths greater than 2000 Å (Khare and Sagan, 1971; Sagan and Khare, 1971) is also a low-yield synthesis, but the amount of energy is much greater in this region of the sun's spectrum. In this synthesis, only H_2S absorbs the ultraviolet light, but the photodissociation of H_2S produces a hydrogen atom having a high kinetic energy, which activates or dissociates the methane, ammonia, and water.

A photochemical source of HCN using very short wavelengths (796 to 912 Å) has been proposed by Zahnle (1986). The nitrogen atoms that are produced diffuse lower into the atmosphere and react with CH_2 and CH_3, producing HCN. The process depends on the N atoms reacting with nothing else before they encounter the CH_2 and CH_3. Whether this proposal is valid remains to be determined.

The most widely used sources of energy for laboratory syntheses of prebiotic compounds are electric discharges. These include sparks, semicorona, arc, and silent discharges, with the spark being the most frequently used type. The ease of handling and high efficiency of electric discharges favor their use, but the most important reason is that electric discharges are very efficient in synthesizing hydrogen cyanide, whereas ultraviolet light is not except for the very short wavelengths. Hydrogen cyanide is a central intermediate in prebiotic synthesis; it is needed for amino acid synthesis via the Strecker reaction or via self-polymerization, and most importantly for the prebiotic synthesis of the purines adenine and guanine, two of the nitrogen-containing bases of nucleic acids.

Use of any of these energy sources for prebiotic syntheses requires activation of molecules in a local area, followed by quenching of the activated mixture, and then protection of the synthesized organic compounds from further influence of the energy source. The quenching and protective steps are critical because the organic compounds will be destroyed if they are subjected continuously to the energy source.

CHEMICAL COMPONENTS OF LIVING ORGANISMS

Before discussing the results of various types of prebiotic synthesis, we need to review the types of compounds in present-day living organisms. The chemical reactions within cells are almost all catalyzed by enzymes that are **proteins.** Proteins are also used for other purposes, such as to form connective tissue, but their role as enzymes is the most important. These proteins are made up of 20 different **amino acids.** Thus the prebiotic synthesis of amino acids is a prime objective in studies of the origin of life.

The genetic information in a cell is contained in the nucleic acid DNA (**deoxyribonucleic acid**), encoded by a sequence of four nucleotides. Each of these **nucleotides,** in turn, is composed of a nitrogen-containing base (one of the **purines:** adenine, "A," or guanine, "G," or one of the **pyrimidines:** thymine, "T," or cytosine, "C"), the sugar deoxyribose, and a phosphate. DNA is a helically wound, double-stranded molecule, the two strands linked together by hydrogen bonds between complementary nitrogenous bases ("A" in one strand is bonded to "T" in the other, and "G" in one, bonding with "C" in the other). The information encoded by the sequence of nucleotides in the DNA (commonly referred to as the base sequence) is transcribed to RNA **(ribonucleic acid).** RNA is similar to DNA in that it is composed of nucleotides containing phosphate and three of the same bases as DNA (adenine, guanine, and cytosine); RNA differs from DNA, however, by usually being single-stranded, and by containing the pyrimidine uracil ("U") in place of thymine, and the sugar ribose in place of deoxyribose. The RNA with the instructions for the synthesis of a protein enters a structure within the cell called the **ribosome,** where these instructions are used for the assembly of a protein from activated amino acids.

Membranes are composed of **phospholipids,** which in most cells consist of two fatty acids, a glycerol, phosphate, and choline (or ethanol amine). The energy that drives the biochemical reactions within a cell is stored in molecules of **adenosine triphosphate (ATP).** These molecules consist of adenine, ribose, and three phosphates (that is, the adenine-containing nucleotide of RNA with two extra phosphates, as shown on p. 23). Much of the metabolic system of a cell is devoted to making energy-rich ATP.

A crucial problem for origin of life studies is to establish whether the first organism contained proteins, or RNA, or both, and the question is a matter of debate. In favor of "proteins first" is the fact that amino acids are formed prebiotically and are polymerized more easily than RNA. In favor of "RNA first" is the fact that RNA can carry genetic information, but proteins cannot. Because of the complexities involved, it seems impossible for a protein-synthesizing system using genetic information to have arisen prebiotically all at once. The protein-RNA dilemma has been resolved, at least for the present, in favor of RNA, by the discovery of catalytic (enzymelike) activity in certain RNA molecules known as **ribozymes** (Cech and Bass, 1986). If correct, this interpretation implies that the first organisms were composed of RNA with essentially no role for protein, with this hypothetical earliest biosphere being referred to as the "RNA world" (Gilbert, 1986).

Note, however, that although "RNA first" is the currently prevailing opinion, it is necessary to keep an open mind because so little is known about prebiotic events.

PREBIOTIC SYNTHESIS OF ORGANIC COMPOUNDS

Amino Acids: Synthesis in Strongly Reduced Atmospheres

Mixtures of CH_4, NH_3, and H_2O, with or without added hydrogen (H_2), are considered strongly reducing atmospheres. The atmosphere of Jupiter contains these species with the H_2 in large excess over the CH_4. The first successful prebiotic amino acid synthesis was carried out using an electric discharge as the energy source and a strongly reducing atmosphere (Figure 1.1; Miller, 1953, 1955). The result was a large yield of amino acids (the yield of the amino acid glycine, alone, incorporating 2.1% of the total carbon available), together with hydroxy acids, short aliphatic acids, and urea (Table 1.2). One of the surprising results of this experiment was that the products were not a random mixture of organic compounds; rather, a relatively small number of compounds were produced in substantial yield. In addition, the compounds were, with a few exceptions, of biological importance.

The mechanism of synthesis of the amino and hydroxy acids was investigated (Miller, 1957). It was shown that the amino acids were not formed directly in the electric discharge, but were the result of solution reactions of smaller molecules produced in the discharge, in

FIGURE 1.1

Apparatus used for the electric discharge synthesis of amino acids and other organic compounds in a strongly reducing atmosphere of methane (200 torr), ammonia (200 torr), hydrogen (100 torr), and water. The smaller (500 cm³) flask contained about 250 ml of H_2O, and because the temperature near the electrodes was about 65°C, the pH_2O was 190 torr. Results from experiments carried out in this apparatus are summarized in Table 1.2.

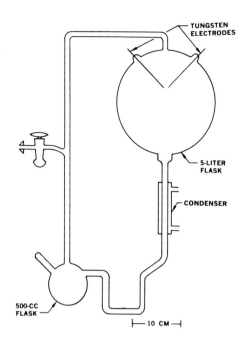

TABLE 1.2

Yields from Sparking a Mixture of CH_4, NH_3, H_2O, and H_2. The Percentage Yields Are Based on Carbon (59 mmoles (710 mg) of Carbon Was Added as CH_4).

Compound	Yield	
	μmoles	%
Glycine	630	2.1
Glycolic acid	560	1.9
Sarcosine	50	0.25
Alanine	340	1.7
Lactic acid	310	1.6
N-Methylalanine	10	0.07
α-Amino-n-butyric acid	50	0.34
α-Aminoisobutyric acid	1	0.007
α-Hydroxybutyric acid	50	0.34
β-Alanine	150	0.76
Succinic acid	40	0.27
Aspartic acid	4	0.024
Glutamic acid	6	0.051
Iminodiacetic acid	55	0.37
Iminoacetic-propionic acid	15	0.13
Formic acid	2330	4.0
Acetic acid	150	0.51
Propionic acid	130	0.66
Urea	20	0.034
N-Methyl urea	15	0.051
Total		15.2

particular, reactions of hydrogen cyanide and aldehydes (CHO-containing compounds). The reactions are summarized as follows (in which "R" represents the side chain of the aldehydes or amino acids):

$$\text{RCHO} + \text{HCN} + \text{NH}_3 \rightleftharpoons \underset{\text{amino nitrile}}{\text{RCH(NH}_2)\text{CN}} \xrightarrow{\text{H}_2\text{O}}$$
$$\underset{\text{aldehyde}}{}$$

$$\underset{}{\text{RCH(NH}_2)\overset{\overset{\text{O}}{\|}}{\text{C}}-\text{NH}_2} \xrightarrow{\text{H}_2\text{O}} \underset{\text{amino acid}}{\text{RCH(NH}_2)\text{COOH}}$$

$$\underset{\text{aldehyde}}{\text{RCHO} + \text{HCN}} \rightleftharpoons \underset{\text{hydroxy nitrile}}{\text{RCH(OH)CN}} \xrightarrow{\text{H}_2\text{O}} \text{RCH(OH)}\overset{\overset{\text{O}}{\|}}{\text{C}}-\text{NH}_2 \xrightarrow{\text{H}_2\text{O}} \underset{\text{hydroxy acid}}{\text{RCH(OH)COOH}}$$

These reactions were later studied in detail, and their equilibrium and rate constants were measured (Miller and Van Trump, 1981). The results show that amino and hydroxy acids can be synthesized at high dilutions of HCN and aldehydes in a simulated primitive ocean. Note also that the rates of these reactions are rather rapid on the geological time scale; the half-lives for the hydrolysis of the intermediate products in the reactions, amino and hydroxy nitriles, are less than a thousand years at 0°C.

This synthesis of amino acids, called the **Strecker synthesis**, requires the presence of the ammonium ion, NH_4^+ (and therefore, of NH_3) in the primitive environment. On the basis of the experimentally determined equilibrium and rate constants, it can be shown (Miller and Van Trump, 1981) that equal amounts of amino and hydroxy acids are obtained when the NH_4^+ concentration in the simulated primitive ocean is about 0.01 M at pH 8 and 25°C, the NH_4^+ concentration being insensitive to temperature and pH. This translates into a low partial pressure of NH_3 in the atmosphere (2×10^{-7} atm at 0°C, and 4×10^{-6} atm at 25°C). Thus at least a small amount of atmospheric NH_3 would seem necessary for amino acid synthesis. A similar estimate of the NH_4^+ concentration in the primitive ocean can be obtained from the equilibrium decomposition of aspartic acid, a prebiotically produced amino acid (Bada and Miller, 1968). Ammonia would have been decomposed in the early environment by ultraviolet light, but mechanisms for its resynthesis are also known. The details of the ammonia balance on the primitive earth remain to be worked out.

In a typical electric discharge experiment, the partial pressure of CH_4 is 0.1 to 0.2 atm (i.e., 100 to 200 torr). This pressure is used for convenience, and it is likely (but has never been demonstrated) that organic compound synthesis would work at much lower partial pressures of methane. There are no estimates available for $p\text{CH}_4$ on the primitive earth, but low levels (10^{-5} to 10^{-3} atm) seem plausible. Higher pressures are not reasonable because the sources of energy would convert the CH_4 to organic compounds too rapidly for higher pressures of CH_4 to build up.

Ultraviolet light acting on a strongly reducing mixture of gases is not efficient in producing amino acids, as discussed earlier. Heating reactions **(pyrolysis)** of CH_4 and NH_3 give very low yields of amino acids. The pyrolysis conditions are from 800° to 1200°C with contact times of a second or less (Lawless and Boynton, 1973). However, the pyrolysis of CH_4 and other hydrocarbons gives good yields of benzene (C_6H_6), phenylacetylene (C_8H_6), and many other hydrocarbons. It can be shown that phenylacetylene would be converted to the amino acids phenylalanine and tyrosine in the primitive ocean (Friedmann and Miller, 1969). Pyrolysis of hydrocarbons in the presence of NH_3 gives substantial yields of indole ($\text{C}_8\text{H}_7\text{N}$), which can be converted to the amino acid tryptophan in the primitive ocean (Friedmann et al., 1971).

Because NH_3 would have dissolved in the ocean, rather than accumulating in large amounts in the early atmosphere, a mixture of CH_4, N_2, H_2O, and traces of NH_3 is a more realistic atmosphere for the primitive earth. Such an atmosphere, however, would nevertheless be strongly reducing. Experimental studies (Figure 1.2) have demonstrated that this mixture of gases is quite effective with an electric discharge in producing amino acids (Ring et al., 1972; Wolman et al., 1972). The yields are somewhat lower than with higher partial pressures of NH_3, but the products are more diverse (Table 1.3). Hydroxy acids, short aliphatic acids, and dicarboxylic acids are produced along with the amino acids (Peltzer and Bada, 1978; Peltzer et al., 1984). Ten of the 20 amino acids that occur in the proteins of present-day organisms are produced directly in this experiment. With the addition of the amino acids asparagine and glutamine, which are formed in the experiment but are hydrolyzed before analysis, and methionine, which is formed when H_2S is added to the reaction mixture (Van Trump and Miller, 1972), we can say that 13 of the 20 amino acids occurring in modern proteins can be formed in this single experiment. The amino acid cysteine has been produced prebiotically via photolysis of CH_4, NH_3, H_2O, and H_2S (Khare and Sagan, 1971). The pyrolysis of hydrocarbons, as discussed earlier, leads to the amino acids phenylalanine, tyrosine, and tryptophan. This leaves only the three basic amino acids, lysine, arginine, and histidine, unaccounted for. Although there are, so far, no established prebiotic syntheses of these amino acids, there is no fundamental reason why these compounds cannot be synthesized, and their prebiotic synthesis may be accomplished before too long.

Amino Acids: Synthesis in Mildly Reducing and Nonreducing Atmospheres

There has been less experimental work with gas mixtures containing CO and CO_2 as carbon sources instead of CH_4. Spark discharges have been the source of energy most extensively investigated (Abelson, 1965; Schlesinger and Miller, 1983a,b; Stribling and Miller, 1987). Figure 1.3 compares amino acid yields using CH_4, CO, or CO_2 as a carbon source in the presence of various amounts of H_2. Separate experiments were performed with and without

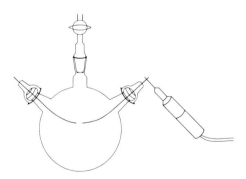

FIGURE 1.2

Apparatus used for the electric discharge synthesis on amino acids in a strongly reducing atmosphere containing traces of ammonia. The 3-liter flask is shown with the two tungsten electrodes and a spark generator. The second electrode is usually not grounded. In the experiments yielding the results summarized in Table 1.3, the flask contained 100 ml of 0.05 M NH_4Cl brought to pH 8.7, giving pNH_3 of 0.2 torr. The pCH_4 was 200 torr and pN_2 was 80 torr. Because the temperature was about 30°C during the 48 hours of sparking, pH_2O was 32 torr.

TABLE 1.3	Yields from Sparking CH_4 (336 mmoles), N_2, and H_2O, with Traces of NH_3.*

	μmoles
Glycine	440
Alanine	790
α-Amino-n-butyric acid	270
α-Aminoisobutyric acid	~30
Valine	19.5
Norvaline	61
Isovaline	~5
Leucine	11.3
Isoleucine	4.8
Alloisoleucine	5.1
Norleucine	6.0
$tert$-Leucine	<0.02
Proline	1.5
Aspartic acid	34
Glutamic acid	7.7
Serine	5.0
Threonine	~0.8
Allothreonine	~0.8
α,γ-Diaminobutyric acid	33
α-Hydroxy-γ-aminobutyric acid	74
α,β-Diaminopropionic	6.4
Isoserine	5.5
Sarcosine	55
N-Ethylglycine	30
N-Propylglycine	~2
N-Isopropylglycine	~2
N-Methylalanine	~15
N-Ethylalanine	<0.2
β-Alanine	18.8
β-Amino-n-butyric acid	~0.3
β-Aminoisobutyric acid	~0.3
γ-Aminobutyric acid	2.4
N-Methyl-β-alanine	~5
N-Ethyl-β-alanine	~2
Pipecolic acid	0.05

* Yield based on the carbon added as CH_4. Glycine = 0.26%; Alanine = 0.71%; total yield of amino acids in the table = 1.90%.

FIGURE 1.3

Amino acid yields based on initial carbon. In all experiments, pN_2 was 100 torr, and pCH_4, pCO, or pCO_2 was 100 torr. The flask contained 100 ml H_2O for the curves with N_2 but no NH_3, and it contained 100 ml of 0.05 M NH_4Cl for the curves with $N_2 + NH_3$ (0.2 torr). The flask was kept at room temperature, and the spark generator was operated continuously for 48 hours.

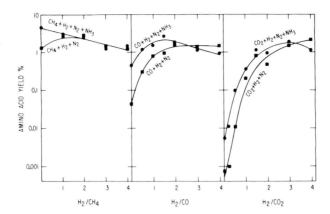

added NH_3. It is clear from the figure that CH_4 is the best source of amino acids using a spark discharge, but that CO and CO_2 are almost as good if a high H_2 to C ratio is used. Without added H_2, however, the amino acid yields are very low, especially with CO_2 as the carbon source. The amino acids produced in the CH_4 experiments are similar to those shown in Table 1.3. With CO and CO_2, glycine was the predominant amino acid with little else besides some alanine being produced.

The implications of these results for the composition of the atmosphere of the primitive earth is that CH_4 is the best carbon source for prebiotic synthesis, especially for amino acid synthesis. Although glycine was essentially the only amino acid synthesized in the spark discharge experiments with CO and CO_2, other amino acids (e.g., serine, aspartic acid, alanine) would probably have been formed from this glycine, H_2CO (formaldehyde), and HCN as the primitive ocean matured. Since we do not know which amino acids made up the first organism, we can only say that CO and CO_2 are less favorable than CH_4 for amino acid synthesis, but that the amino acids produced from CO and CO_2 may have been adequate. The synthesis of purines and sugars we describe later would not be greatly different with CH_4, CO, or CO_2 as long as sufficient H_2 was available. Although the spark discharge yields of amino acids, HCN, and H_2CO are about the same with CH_4, with $H_2/CO > 1$, and with $H_2/CO_2 > 2$, for the latter two reaction mixtures it is not clear how such high molecular hydrogen to carbon ratios could have been maintained in the primitive atmosphere, since H_2 escapes gravitationally from the earth's atmosphere into outer space. These problems are poorly understood and are beyond the scope of this brief review.

Purines and Pyrimidines

Hydrogen cyanide plays an important role in the synthesis of purines as well as of amino acids. This is illustrated in a remarkable synthesis of the purine adenine. If concentrated solutions of ammonium cyanide are refluxed for a few days, adenine is obtained in up to 0.5% yield along with 4-aminoimidazole-5-carboxamide and the usual cyanide polymer (Oro and Kimball, 1961, 1962).

The mechanism of adenine synthesis in these experiments is probably the following:

$$H-C\equiv N + {}^-C\equiv N \longrightarrow H-\overset{\overset{\displaystyle NH}{\|}}{C}-C\equiv N \xrightarrow{HCN} N\equiv C-\overset{\overset{\displaystyle NH_2}{|}}{C}H-C\equiv N \xrightarrow{HCN}$$

Hydrogen   Cyanide   Aminomalonitrile
cyanide   (HCN trimer)

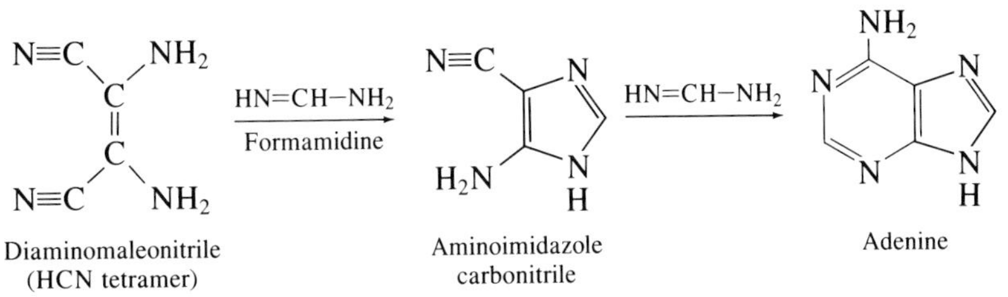

Diaminomaleonitrile (HCN tetramer) → [HN=CH—NH₂, Formamidine] → Aminoimidazole carbonitrile → [HN=CH—NH₂] → Adenine

The difficult step in the synthesis of adenine just summarized is the reaction of the HCN tetramer with formamidine. This step may be bypassed by the photochemical rearrangement of HCN tetramer to aminoimidazole carbonitrile as shown here, a reaction that proceeds readily in present-day sunlight (Sanchez et al., 1967, 1968):

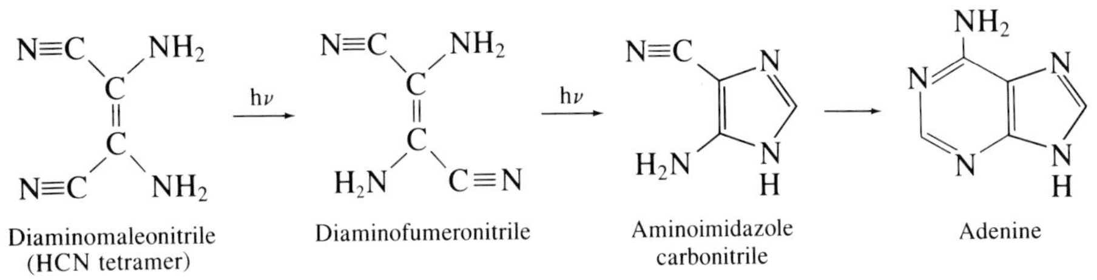

Diaminomaleonitrile (HCN tetramer) → [$h\nu$] → Diaminofumeronitrile → [$h\nu$] → Aminoimidazole carbonitrile → Adenine

A further possibility is that tetramer formation may have occurred in an eutectic solution (i.e., at the lowest melting point possible). High yield of tetramer (>10%) can be obtained by cooling dilute cyanide solutions to between $-10°C$ and $-30°C$ for a few months (Sanchez et al., 1966a).

Guanine and the additional purines, hypoxanthine, xanthine, and diaminopurine could have been synthesized in the primitive environment by variations of the adenine synthesis, as shown here, using aminoimidazole carbonitrile and aminoimidazole carboxamide:

Aminoimidazole carbonitrile → [H_2O] → Aminoimidazole carboxamide

HCN → Adenine; C_2N_2 → Diaminopurine; NCO⁻ → Isoguanine

HCN → Hypoxanthine; C_2N_2 → Guanine; HCO⁻ → Xanthine

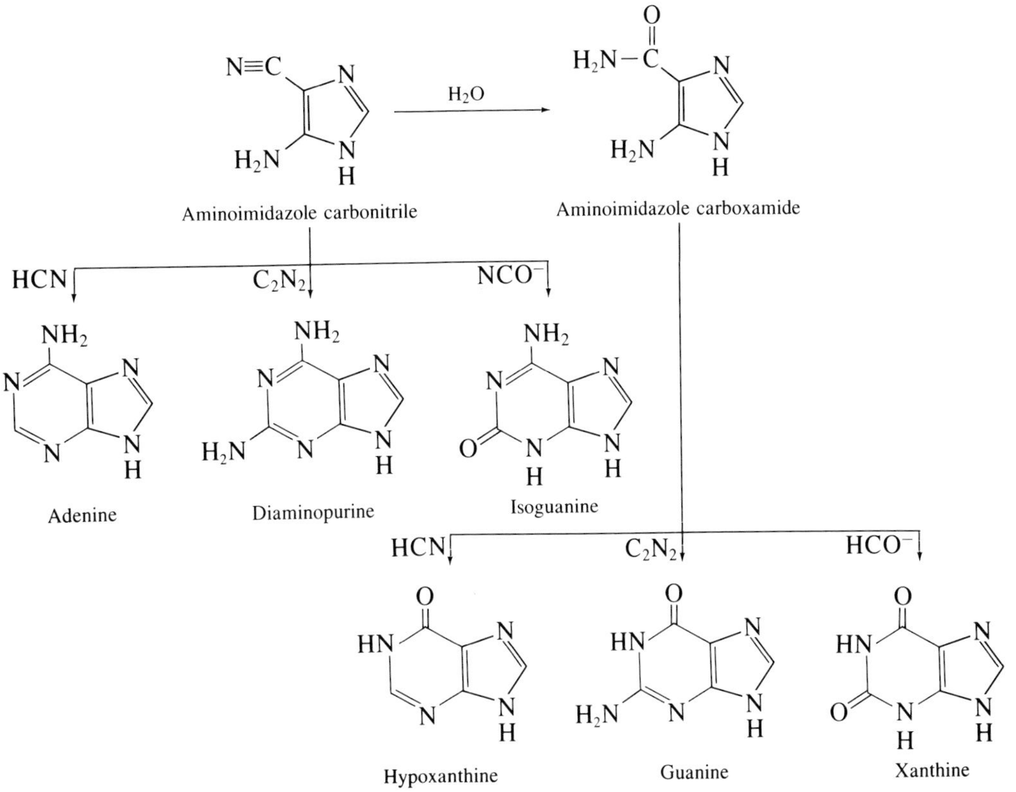

The prebiotic synthesis of the pyrimidines of nucleic acids involves cyanoacetylene, which is synthesized in good yield by sparking mixtures of CH_4 and N_2. As shown here, cyanoacetylene reacts with cyanate (NCO^-) to give the pyrimidine cytosine (Ferris et al., 1968; Sanchez et al., 1966b):

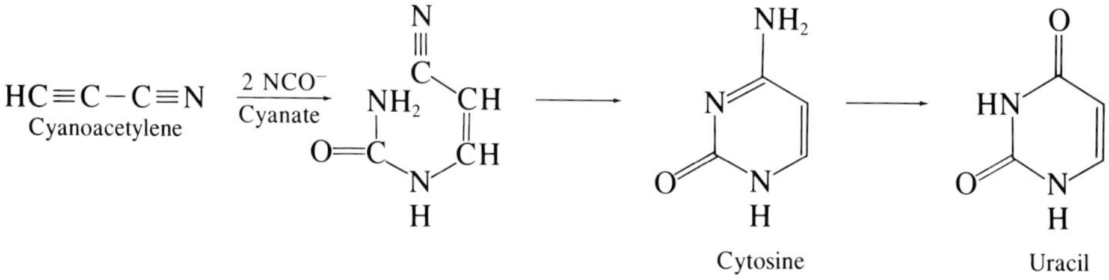

As shown, the cytosine, in turn, can be converted to uracil. Cyanate can come from cyanogen (NCCN) or by the decomposition of urea (H_2NCONH_2), compounds that are both produced readily in prebiotic syntheses.

An alternative prebiotic synthesis of the pyrimidines cytosine and uracil starts with cyanoacetaldehyde, derived from the hydration of cyanoacetylene, which reacts with guanidine to give diaminopyrimidine. This is then hydrolyzed to cytosine and uracil (Ferris et al., 1974):

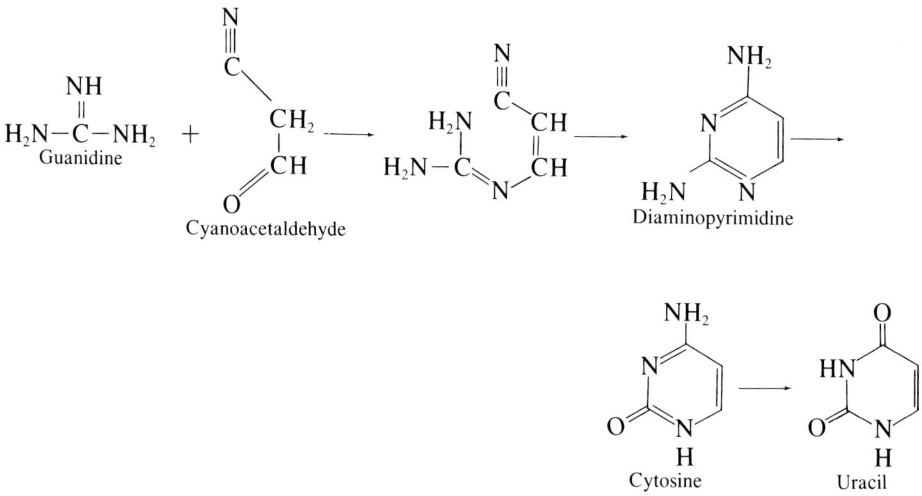

Sugars

The synthesis of sugars from formaldehyde under alkaline conditions was discovered long ago (Butlerow, 1861). However, the Butlerow synthesis, or **"formose reaction,"** is very complex and incompletely understood. It depends on the presence of a suitable inorganic catalyst, with calcium hydroxide [$Ca(OH)_2$] and calcium carbonate ($CaCO_3$) being the most commonly used. In the absence of such mineral catalysts, little or no sugar is obtained. At 100°C, clays such as kaolin serve to catalyze formation of simple sugars, including ribose (the sugar in RNA), in small yields from dilute (0.01 M) solutions of formaldehyde (Gabel and Ponnamperuma, 1967; Reid and Orgel, 1967).

The reaction is autocatalytic and proceeds in a series of stages through glycolaldehyde, glyceraldehyde and dihydroxyacetone, four-carbon sugars, and five-carbon sugars to give finally hexoses, six-carbon sugars, including the biologically important sugars glucose and fructose. One proposed reaction sequence is the following:

REVERSE ALDOL

$$CH_2O \xrightarrow{CH_2O} \begin{array}{c} CHO \\ | \\ CH_2OH \end{array} \xrightarrow{CH_2O} \begin{array}{c} CHO \\ | \\ CHOH \\ | \\ CH_2OH \end{array} \rightleftharpoons \begin{array}{c} CH_2OH \\ | \\ C=O \\ | \\ CH_2OH \end{array} \xrightarrow{CH_2O} \begin{array}{c} CH_2OH \\ | \\ C=O \\ | \\ CHOH \\ | \\ CH_2OH \end{array} \rightleftharpoons \begin{array}{c} CHO \\ | \\ CHOH \\ | \\ CHOH \\ | \\ CH_2OH \end{array}$$

Formaldehyde   Glycolaldehyde   Glyceraldehyde Dihydroxyacetone

PENTOSE   **HEXOSE**
(Five-carbon)   (Six-carbon)
sugars   sugars

There are two problems with the formose reaction as a source of sugars on the primitive earth. The first is the instability of sugars; they decompose in a few hundred years or less at 25°C. There are a number of possible ways to stabilize sugars; the most interesting is to convert the sugar to a glycoside of a purine or pyrimidine (i.e., attaching the sugar to a purine or pyrimidine from which it can be released subsequently by hydrolysis). The second problem is that the formose reaction gives a wide variety of sugars, both straight chain and branched. Indeed, more than 40 different sugars have been separated from one reaction mixture (Decker et al., 1982). Ribose occurs in this mixture, but it is not particularly abundant. It is difficult to envision how the relative yield of ribose, needed for formation of RNA, could be greatly increased in this reaction, or how any prebiotic reaction producing sugars could give mostly ribose. It therefore has become apparent that **ribonucleotides,** ribose-containing nucleotides, could not have been the first components of prebiotic nucleic acids (Shapiro, 1988). A number of ribose substitutes have been proposed (Joyce et al., 1987), which include the two compounds on the right:

HOCH₂ O Base   HOCH₂ O Base   HOCH₂   Base

OH   OH   OH   OH

Riboside   "Ribose substitute"   "Ribose substitute"
(Ribose bonded   bonded to a   bonded to a
to a nitrogenous   nitrogenous base   nitrogenous base
base)

There are many other possible substitutes for ribose, but the ones shown are attractive because they are open chain, flexible molecules that do not have an asymmetric carbon. The prebiotic synthesis of these compounds has not yet been demonstrated.

Other Prebiotic Compounds

In addition to the foregoing, numerous other compounds have been synthesized under primitive earth conditions, including the following:

Dicarboxylic acids

Tricarboxlic acids

Fatty acids C_2-C_{10} (branched and straight)

Fatty alcohols (straight chain via Fischer-Tropsch reaction)

Porphin

Nicotinonitrile and nicotinamide

Triazines

Imidazoles

Other prebiotic compounds that may have been involved in polymerization reactions include the following:

Cyanate [NCO^-]

Cyanamide [H_2NCN]

Cyanamide dimer [$H_2NC(NH)NH-CN$]

Dicyanamide [$NC-NH-CN$]

Cyanogen [$NC-CN$]

HCN tetramer

Diiminosuccinonitrile

Acylthioesters

Phosphate polymers

Production Rates of Organic Compounds on the Primitive Earth

A definitive statement cannot be made about the production rates and the concentrations of compounds in the primitive ocean because the atmospheric composition, ambient temperature, and ocean size are unknown. In addition, the mechanisms of destruction of organic compounds in the early environment are poorly understood. Nevertheless, quantitative estimates have been made (Stribling and Miller, 1987). The calculations are complicated and uncertain, but the results give an amino acid concentration of 3×10^{-4} M and an adenine concentration of about 1.5×10^{-5} M. This can be considered as a relatively concentrated prebiotic soup.

Compounds That Have Not Been Synthesized Prebiotically

It is a matter of opinion as to what constitutes a plausible prebiotic synthesis. In some syntheses, the conditions are so forced (e.g., by use of anhydrous solvents) or the concentrations are so high (e.g., 10 M formaldehyde) that the syntheses could not be expected to have occurred extensively (if at all) on the primitive earth. Reactions under these and other extreme conditions cannot be considered plausibly prebiotic.

There have been many reported "prebiotic syntheses" in which the compound claimed to have been produced has not been properly identified. At present, the best method for unequivocal identification is gas chromatography-mass spectrometry of a suitable chemically synthesized derivative, although melting points and mixed melting points can sometimes be used. Identification of a compound cannot be established unequivocally based solely on the use of an amino acid analyzer, or via chromatography in multiple solvent systems.

Some important biological compounds that do not yet have adequate prebiotic syntheses are the following:

Arginine	Thiamine
Lysine	Riboflavin
Histidine	Folic acid
Straight chain fatty acids	Lipoic acid
Porphyrins	Biotin
Pyridoxal	

It is probable that plausible prebiotic syntheses will become available before too long for some of these compounds. In other cases, the compounds may not have been synthesized prebiotically because their occurrence in living systems is a result of intracellular biochemical evolution that occurred after the origin of life.

EXTRATERRESTRIAL ORGANIC SYNTHESES

Organic Compounds in Carbonaceous Chondrites

Prebiotic experiments under controlled laboratory conditions can be criticized as not representing realistic geological conditions. This criticism is difficult to answer directly because no rocks are known to have survived to the present from the earliest several hundred million years of earth history (see Chapter 2). However, organic compounds are abundant in certain carbonaceous meteorites, in the atmospheres of Jupiter and Saturn as well as Titan, the largest satellite of Saturn, and in interstellar dust clouds. Studies of these extraterrestrial organic materials establish that the synthesis of organic compounds, in the absence of life, occurs widely in the solar system and beyond.

Most stony meteorites do not contain much carbon, but there is a group of meteorites called **carbonaceous chondrites** that contain 0.5% to as much as 5% organic carbon. The term *chondrite* refers to the occurrence of millimeter-size spheres called chondrules that are thought to have formed from molten silicate drops in the gaseous cloud (the solar nebula) that condensed to form the solar system. Although the occurrence of organic compounds in these meteorites has been long accepted, until recent years attempts to analyze for specific compounds failed because of inadequate analytical methods and because these meteorites are easily contaminated by terrestrial organic compounds in the air and by bacteria.

These problems were overcome when a carbonaceous chondrite fell on 28 September 1969 near Murchison, Australia, an organic-rich meteorite that is therefore called the Murchison meteorite, or simply "Murchison" (Figure 1.4). Numerous pieces of the meteorite were picked up immediately, and use of the most reliable analytical methods (gas chromatography-mass spectrometry) revealed surprisingly large amounts of amino acids (Kvenvolden et al., 1970, 1971; Oro et al., 1971). The first report identified seven amino acids (glycine, alanine, valine, proline, glutamic acid, sarcosine, and α-aminoisobutyric acid), of which all but valine and proline had also been found in the original prebiotic electric discharge experiments (Miller, 1953, 1955, 1977). A second report identified 18 amino acids in the meteorite, of which 9 had previously been identified in the original electric discharge experiments. At that time, only the hydrophobic amino acids from the electric discharge experiments (Table 1.2) had been identified firmly, and so the discharge products were further examined in an attempt to detect the nine additional (chiefly

FIGURE 1.4

A fragment of the Murchison carbonaceous chondrite (specimen LC 848 from the UCLA collections), a chondritic meteorite containing 2.5% organic matter, collected after an observed meteorite fall on 28 September 1969 near Murchison, Victoria, Australia. The smooth, charred exterior fusion crust on this specimen was produced by frictional heating as the meteorite passed through the earth's atmosphere. Prominent glassy, granular condrules give the interior of the meteorite a speckled appearance. (Photograph courtesy of J. W. Schopf.)

nonprotein) amino acids found in Murchison. All of these amino acids were subsequently detected and identified (Ring et al., 1972; Wolman et al., 1972).

There is a striking similarity, in terms of both the compounds identified and of their relative abundances, between the amino acids produced by electric discharge and those occurring in the Murchison meteorite. Table 1.4 compares the results of the two sets of analyses. The most notable difference between the two is the occurrence of pipecolic acid in the Murchison, an amino acid produced in very low yield in the electric discharge. Proline is also present in substantially greater abundance in Murchison than in the electric discharge products. The amount of α-aminoisobutyric acid is greater than α-amino-n-butyric acid in the meteorite, but the reverse is observed in products of the electric discharge. However, these differences in relative abundances of amino acids do not detract from the overall picture. Indeed, the ratio of α-aminoisobutyric acid to glycine has been shown to be quite different in two meteorites of the same type (0.4 in Murchison and 3.8 in Murray; Cronin and Moore, 1971). A similar comparison has been made between the dicarboxylic acids in Murchison (Lawless et al., 1974) and those produced by an electric discharge (Zeitman et al., 1974), and the product ratios are quite similar to those observed for the amino acids.

The close correspondence between the amino acids found in the Murchison meteorite and those produced by an electric discharge synthesis, both as to the amino acids produced and their relative abundance, suggests that the amino acids in the meteorite were synthesized on the *parent body* of the meteorite (the larger body from which the meteorite was derived) by means of an electric discharge or some analogous processes. A quantitative comparison of the amino acid and hydroxy acid abundances (Peltzer and Bada, 1978) shows that these compounds can be accounted for by a particular type of synthetic reaction, a Strecker-Cyanohydrin synthesis, occurring on the parent body (Peltzer et al., 1984). Electric discharges appear to be the most favored source of energy for such a synthesis, but sufficient data are not available to make a realistic comparison with other energy sources.

These results show that a prebiotic synthesis of organic compounds took place on the parent body of Murchison, generally thought to have been as asteroid. This does not prove that such syntheses took place on the primitive earth, but it does make such proposals quite plausible.

Interstellar Molecules

In the past 20 years a large number of organic molecules have been identified in interstellar dust clouds, mostly based on emission lines observed in the microwave region of the spectrum (for a summary, see Mann and Williams, 1980). The concentration of these molecules is very low (less than a few molecules per cm^3), but because interstellar dust clouds are so vast, the total number of organic molecules in a dust cloud is large. The molecules identified include formaldehyde (H_2CO), hydrogen cyanide (HCN), acetaldehyde (C_2H_4O), and cyanoacetylene (HC_3N). These are important compounds in prebiotic synthesis, and this immediately raises question of whether such interstellar molecules may have played a role in the origin of life on earth. In order for this to have taken place, it would have been necessary for the molecules to have been greatly concentrated in the solar nebula and to have arrived on earth without having been destroyed by ultraviolet light or pyrolysis. This appears unlikely. In addition, it would have been necessary for some of these molecules to have been continuously synthesized on the primitive earth (unless life started

TABLE 1.4		

Relative Abundances of Amino Acids in the Murchison Meteorite and in an Electric Discharge Synthesis.†

Amino acid	Murchison meteorite	Electric discharge
Glycine	****	****
Alanine	****	****
α-Amino-n-butyric acid	***	****
α-Aminoisobutyric acid	****	**
Valine	***	**
Norvaline	***	***
Isovaline	**	**
Proline	***	*
Pipecolic acid	*	<*
Aspartic acid	***	***
Glutamic acid	***	**
β-Alanine	**	**
β-Amino-n-butyric acid	*	*
β-Aminoisobutyric acid	*	*
γ-Aminobutyric acid	*	**
Sarcosine	**	***
N-Ethylglycine	**	***
N-Methylalanine	**	**

†Mole ratio to glycine (=100): * 0.05–0.5; ** 0.5–5; *** 5–50; **** >50.

very quickly) because of their inherent instability, and an interstellar source could not have accomplished this synthesis.

For these reasons, it is generally felt that interstellar organic molecules played, at most, a minor role in the origin of life on earth. However, the presence of so many molecules of prebiotic importance in interstellar space, combined with the fact that their synthesis must differ from that on the primitive earth where the conditions were vastly different, indicates that some organic molecules are particularly easily synthesized in the absence of living systems. In other words, there appears to be a universal organic chemistry, one that is manifest in interstellar space, occurs in the atmospheres of the major planets of the solar system, and must also have occurred in the reducing atmosphere of the primitive earth.

Input from Comets and Meteorites

It is generally accepted that impacts of comets and meteorites have been important in the earth's history, with the most recent large impact occurring 65 Ma ago and possibly playing a role in the Cretaceous-Tertiary extinction event (see Chapter 5). As we discussed, impacts were much more frequent earlier than 3,800 Ma ago and may have influenced the timing of the origin of life. Our question here is whether these objects were a source of organic compounds for the origin of life. In other words, how much organic material came in with the colliding objects, and what portion of it survived impact? Some meteorite-type amino acids (in particular, α-aminoisobutyric and isovaline) have been found at the Cretaceous-Tertiary boundary in Denmark (Zhao and Bada, 1989), but the exact source of these amino acids is uncertain, and survival of organic compounds brought to earth on relatively large objects is thought to be difficult. Cosmic dust particles ($<10\mu$m diameter) can survive impact because the atmosphere slows them down. Organic compounds in meteorites of less than a few meters diameter also survive impact, but larger objects are heated to a high temperature (Anders, 1989). Another source of organic compounds would be comets, but the organic compounds in comets would survive only if the earth were to have a very dense atmosphere (about ten times that of the present), sufficient to brake the comet's impact (Chyba et al., 1990).

Production of organic compounds on the primitive earth under reducing conditions would far exceed the inputs from cosmic dust, meteorites, and comets. Only if conditions were nonreducing (e.g., an atmosphere of CO_2, N_2, and H_2O with very little H_2) would extraterrestrial inputs of organic compounds be quantitatively important.

POLYMERIZATION PROCESSES ON THE PRIMITIVE EARTH

We have established that numerous biologically important types of small molecules (**monomers**) can be synthesized under plausible prebiotic conditions for the primitive earth, and that similar syntheses have occurred both elsewhere in the solar system and in interstellar space. Living systems, however, are composed of larger molecules, **polymeric** aggregates of such monomeric components. A crucial question with regard to the origin of life on earth is therefore, "How might the prebiotic monomers have become chemically combined to form the type of polymers of which living systems are composed?"

Polymerization of amino acids into proteins and of nucleotides into DNA and RNA is central to the metabolism of living organisms, and a large part of the energy produced by an organism is expended in these processes. These reactions, which are **dehydration condensation reactions** (the process of removal of H_2O from, and the linkage of, adjacent monomers), are thermodynamically unfavorable, and ATP is used in organisms to overcome the thermodynamic barrier. In prebiotic processes this free energy barrier can be overcome in two ways. The first is to drive the dehydration reaction (removal of water) by coupling it to the **hydration** (addition of water) of a higher energy compound. The second method is to remove the water by heating. In principle, visible or ultraviolet light could drive these reactions, but so far no one has demonstrated such prebiotic processes.

Peptide Synthesis

Proteins (including enzymes) are linear polymers composed of amino acid monomers. The component amino acids, in turn, are linked one to another by the **peptide bond,** OC—NH, and proteins are therefore referred to as polypeptides. Thus a principal problem for the prebiotic synthesis of protein-type molecules is the formation of peptide bonds. An example of the use of a high-energy compound to form a peptide bond between two amino acids is the use of cyanamide, which is hydrated to urea, resulting in the linkage of glycine to leucine via a dehydration condensation reaction and formation of a dipeptide:

$$H_3N^+-CH_2-\overset{\overset{\displaystyle O}{\|}}{C}-O^- \; + \; H_3N^+-\overset{\overset{\displaystyle R}{|}}{CH}-\overset{\overset{\displaystyle O}{\|}}{C}-O^- \; + \; H_2N-C\equiv N \longrightarrow H_3N^+-CH_2-\overset{\overset{\displaystyle O}{\|}}{C}-NH-\overset{\overset{\displaystyle R}{|}}{CH}-COO^- \; +$$

Glycine   Leucine   Cyanamide   Glycylleucine

$$H_3N^+-\overset{\overset{\displaystyle R}{|}}{CH}-\overset{\overset{\displaystyle O}{\|}}{C}-NH-CH_2-\overset{\overset{\displaystyle O}{\|}}{C}-O^- \; + \; H_2N-\overset{\overset{\displaystyle O}{\|}}{C}-NH_2$$

Leucylglycine   Urea

The yields of this reaction are at best a few percent (Ponnamperuma and Peterson, 1965). Similar results of comparable low yield have been obtained using other reagents (for a review, see Hulshof and Ponnamperuma, 1976). Use of phosphates and ATP gives somewhat higher yields of dipeptides, but these phosphates can also react with the amino acid nitrogen and thereby inhibit peptide synthesis.

One way around this problem is to use imidazole as a catalyst (Lohrmann and Orgel, 1973; Weber and Orgel, 1978). The ATP reacts with the amino (NH_2) group of the amino acid forming the phosphoroamidate which, even if heated, does not polymerize. In the presence of imidazol (Im), the phosphoroamidate of the amino acid is converted to the imidazolide (ImPA):

$$ATP + Glycine \longrightarrow {}^-OOC-CH_2-NH-\overset{\overset{\displaystyle O}{\|}}{\underset{\underset{\displaystyle {}^-O}{|}}{P}}-O-Ad$$

Phosphoroamidate

Imidazole   heat

$$Glycine + Im-\overset{\overset{\displaystyle O}{\|}}{\underset{\underset{\displaystyle {}^-O}{|}}{P}}-O-Ad \qquad No\ peptide$$

Imidazolide (ImPA)

$$\underset{\text{Imidazole}}{\overset{\displaystyle N \;\; NH}{\bigtriangledown}} \; + \; H_3^+N-CH_2-\overset{\overset{\displaystyle O}{\|}}{C}-O-\overset{\overset{\displaystyle O}{\|}}{\underset{\underset{\displaystyle {}^-O}{|}}{P}}-O-Ad \longrightarrow Oligoglycine\ peptides$$

Aminoacyladenoylate

ImPA then reacts with the amino acid at the carboxyl (COOH) group forming the aminoacyladenoylate, which polymerizes relatively well.

An alternate approach to polypeptide synthesis is to heat mixtures of amino acids at low relative humidities. This process has been investigated extensively by Fox and coworkers (Fox and Dose, 1972). Typical conditions involve using an excess of aspartic and glutamic acids and heating at 150° to 180°C for a few hours:

$$\text{n amino acid} \xrightleftharpoons[\text{excess asp or glu}]{150° \text{ to } 180°C} \text{polypeptide} + \text{n H}_2\text{O}$$

The yields are good ($\sim$50%) and products having molecular weights of a few thousand can be obtained. The problem with this approach is that *dry* temperatures of 150° to 180°C do not occur extensively on the earth. Higher temperatures are available in volcanoes (1200°C), but such temperatures would destroy the amino acids. Hot springs are not suitable for such syntheses because of the excess water. A few experiments with these thermal polymerizations have been done at lower temperatures (Rohlfing, 1976), but the yields are small, and it is not clear whether the results obtained at 150° to 180°C can be obtained over longer periods at the more reasonable temperatures of 50° to 100°C.

Prebiotic Synthesis of Nucleosides and Nucleotides

One of the more difficult prebiotic polymerization reactions is the linkage of a monomeric sugar (e.g., ribose) to a monomeric purine or pyrimidine to form a **nucleoside.** Heating of ribose and purines at 100°C gives fair yields (2% to 20%) of a mixture of isomers of the purine ribosides, with some of the correct β-isomer being produced (Fuller et al., 1972). However, heating a mixture of pyrimidines and ribose gives no detectable yield of nucleosides. Whether the inability to synthesize pyrimidine ribosides is a matter of not using the correct experimental conditions, or whether pyrimidine nucleosides were not present in the first organisms, remains to be determined. It also has to be kept in mind that large amounts of pure ribose are not prebiotic, as discussed earlier with regard to sugar synthesis.

The phosphorylation of nucleosides to form **nucleotides** presents fewer problems. The first step is the synthesis of polyphosphates from ammonium dihydrogen phosphate ($\text{NH}_4\text{H}_2\text{PO}_4$). Dihydrogen phosphates do not occur on the earth at the present time, and even monohydrogen phosphates are rare. The reason for this is that hydroxyl apatite $[\text{Ca}_{10}(\text{PO}_4)_6(\text{OH})_2]$, the material of which vertebrate bone is composed, is the major occurring phosphate mineral, and it is very insoluble. However, the precipitation of apatite is slow, especially in the presence of Mg^{2+} (Handschuh and Orgel, 1972). Phosphate does not build up in the ocean at the present time because organisms remove it. However, on the primitive earth there might have been a buildup of phosphate, and the presence of NH_4^+ in the oceans would aid this. One can then envision the evaporation of seawater in a lagoon, resulting in the precipitation of MgNH_4PO_4 and $\text{NH}_4\text{H}_2\text{PO}_4$. On heating to 80° to 100°C, especially in the presence of urea, polyphosphates are produced in good yield (Handschuh and Orgel, 1972):

$$\text{n NH}_4\text{H}_2\text{PO}_4 \xrightarrow[\text{excess urea}]{85° \text{ to } 100°C} (\text{NH}_4\text{PO}_3)_n + \text{n H}_2\text{O}$$

These polyphosphates can then phosphorylate nucleosides rather efficiently (Lohrmann and Orgel, 1976; Lohrmann et al., 1980). If amino acids are present, the AMP-amino acids are synthesized, but if imidazole is present imidazolides are produced, for example, ImPA (see earlier discussion of peptide synthesis). The imidazolides, such as that shown on the left, which are similar in structure to the ATP (shown on the right), are very reactive compounds that can be used for template polymerizations.

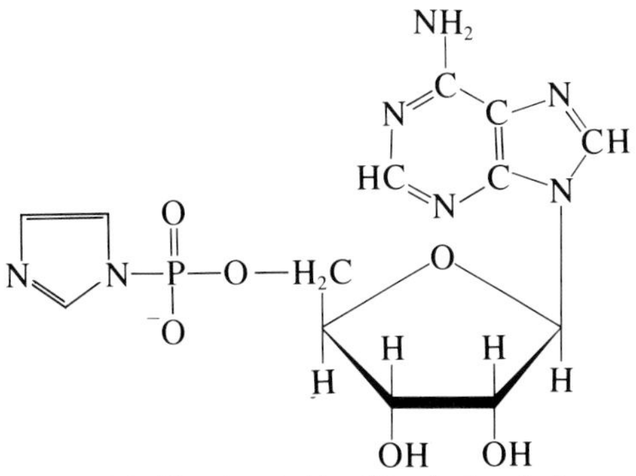

ImPA (Phosphoroimidazolide of adenosine) 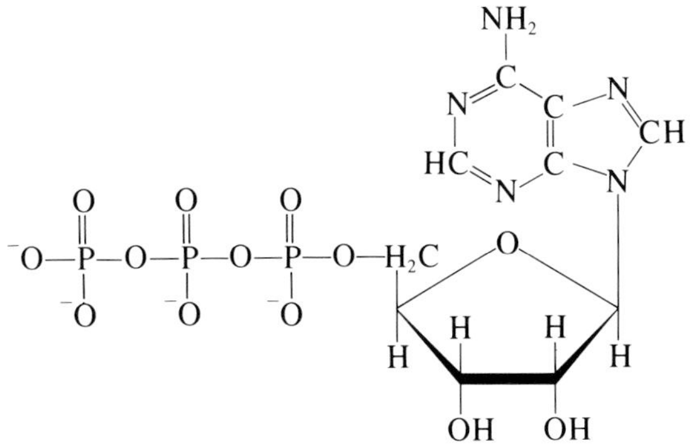   ATP (Adenosine triphosphate)

Imidazole is a reasonable prebiotic compound, since it can be synthesized from glyoxal ($C_2H_2O_2$), NH_3, and H_2CO, or from cyanoacetylene, NH_3, and ultraviolet light. However, it is not clear whether large amounts of imidazole were present in the primitive ocean. It may be that the template polymerizations we describe here were carried out slowly using nucleotide polyphosphates. Such studies are difficult to carry out experimentally because of the long time required.

Template Polymerizations

When the genetic information in DNA is replicated in living systems, the two DNA strands separate, and each strand directs the synthesis of its complement. The actual process is quite complex and is carried out by many enzymes. This process can be called a **template polymerization** because the polymerization (the synthesis of the complementary strand) is directed by the single strand of DNA acting as a template. A major goal of prebiotic research is to find the first version of this process. It must have taken place under reasonable geological conditions without the aid of enzymes.

An extensive investigation of potentially prebiotic template polymerizations has been carried out by Orgel and his co-workers (Lohrmann and Orgel, 1976; Lohrmann et al., 1980). There are a number of interesting results. The reaction of the activated purine ribotides on a polypyrimidine template proceeds quite efficiently, forming oligomers ranging from dimers up to molecules 50 monomeric units long or even longer. Unfortunately, the complementary reaction of activated pyrimidines on a polypurine template does not work (Figure 1.5).

This lack of success is attributed to the inability of pyrimidine nucleosides to stack appropriately on the polypurine (polyadenine or polyguanine) template. It is possible that early template polymerizations used short oligomers (dimers, trimers, etc.) which may have been able to stack, or that there may be conditions not yet found which permit the stacking of pyrimidine nucleosides. Otherwise, it is difficult to imagine that pyrimidines were involved in the earliest genetic system.

A second interesting result from these studies is that 2'-5' phosphodiester bonds between purine ribonucleotides are usually formed more easily than the desired 3'-5' bonds, bonds of the type occurring in the RNA (and DNA) of modern organisms. The 2'-5' bonds hydrolyze more readily than 3'-5' bonds which, given sufficient time, would shift the population of polynucleotides in the primitive ocean over to those having the 3'-5' linkage, but it clearly would be better to obtain initially the 3'-5' isomers. Experimentally, this can

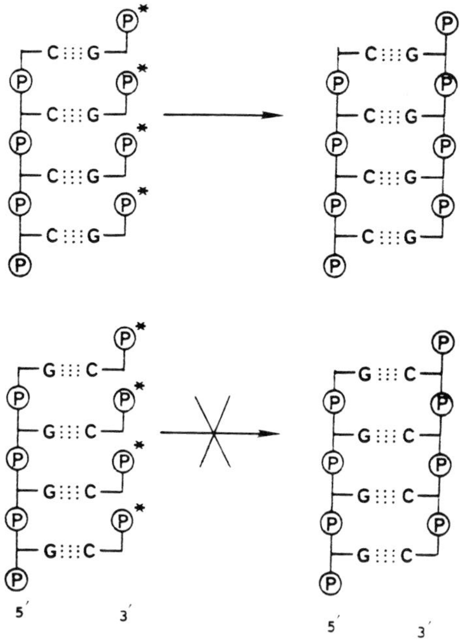

FIGURE 1.5

Schematic representation of template polymerizations discussed in the text. As shown (above), polymerization of an activated purine ribonucleotide (ImP-guanine, "ImPG") on a polypyrimidine template (polycytosine, "PolyC") proceeds efficiently. In contrast (below), an activated pyrimidine ribonucleotide containing cytosine ("ImPC") does not polymerize on a polyguanine template. Not shown is the similar representation of polymerization of ImPA on a PolyU template, and the failure of ImPU to polymerize on a PolyA template.

be accomplished by using a Zn^{2+} catalyst, as long as the Zn^{2+} is not complexed with the buffer salts (Lohrmann et al., 1980).

The most interesting aspect of the Zn^{2+}-catalyzed polymerization is its remarkable fidelity. If equal amounts of ImP-adenine and ImP-guanine are reacted with a polycytosine template, only 1 adenine per 200 guanines is incorporated into the resulting template-produced polymer. This is only an order of magnitude of less fidelity than that exhibited by RNA polymerases occurring in present-day living systems (via which errors are introduced at a frequency of about 1 in 4000).

Limitations of Template Polymerization Syntheses

The fact that activated pyrimidines do not react on a polypurine template, whereas the converse reaction is quite efficient, makes it clear that these experimental template reactions are not efficient enough to replicate genetic information. Attempts to use mixed guanine-cytosine (and adenine-uracil) templates were more successful, but in order for these systems to work the template required about 70% pyrimidines. Therefore, the daughter strand would contain 30% pyrimidines and could not act as a template for further synthesis.

There are additional problems with these types of experiments. The imidazolides of the nucleotides are not realistically prebiotic. In addition, the experiments were conducted using pure ribonucleosides as reagents, whereas prebiotic mixtures would contain purines and pyrimidines bonded to sugars other than ribose. Even more difficult is the requirement in these template polymerizations for the presence of a particular form of pure optically active ribonucleoside (viz., that containing D-ribose), because the presence of L-ribonucleosides inhibits polymerization.

It is clear that several pieces of this polymerization puzzle are missing. The pyrimidines do not seem to be prebiotic (i.e., they do not appear to have played a role in the first informational macromolecules); the requirement for D-ribose also seems implausible; and even phosphate may not have been abundantly present. The only components of RNA that seem very primitive are the purines. It is a real challenge to discover the prebiotic replacements for the pyrimidines and ribose.

SUMMARY AND CONCLUSIONS

The environment of the primitive earth is thought to have been more or less reducing; under these conditions, experimental studies have established that organic compounds would have been synthesized on a large scale. Many of the monomeric compounds thus produced are of importance in present-day organisms. These include amino acids, purines, pyrimidines, and sugars. Syntheses of precursors to such compounds take place at the present time in the atmospheres of Jupiter, Saturn, and Titan, as well as in interstellar space. Similar syntheses took place on the parent body (probably an asteroid) of the carbonaceous chondrites. To date, no highly efficient prebiotically plausible processes for the polymerization of amino acids to form peptides, or of mixtures of purines, pyrimidines, and ribose to form ribonucleosides, have been found. Considerable success has been obtained in the template polymerization of activated purines on polypyrimidine templates, but these reactions have not been accomplished under realistically prebiotic conditions.

Over recent years, investigations of the origin of life on earth have been placed on a sound foundation. Much has been learned, but even more remains to be discovered. As a solvable scientific question of surpassing human interest, the problem of the origin of life stands out as one where the greatest advances are still to be made.

Acknowledgments

This chapter is based on an article originally prepared for a Cold Spring Harbor Symposium on Quantitative Biology (Miller, 1987). The work was supported by NASA Grant NAGW-20.

REFERENCES CITED

Abelson, P. H. 1965. Abiogenic synthesis in the Martian environment. *Proc. Natl. Acad. Sci. USA 54:* 1490–1494.

Anders, E. 1989. Pre-biotic organic matter from comets and asteroids. *Nature 342:* 255–257.

Bada, J. L., and Miller, S. L. 1968. Ammonium ion concentration in the primitive ocean. *Science 159:* 423–425.

Bernal, J. D. 1951. *The Physical Basis of Life* (London: Routledge and Kegan Paul), 80 pp.

Butlerow, A. 1861. Formation synthétique d'une substance sucrée. *C.R. Acad. Sci. 53:*145–147.

Cech, T. R., and Bass, B. L. 1986. Biological catalysis by RNA. *Ann. Rev. Biochem. 55:* 599–629.

Chyba, C. F., Thomas, P. J., Brookshaw, L., and Sagan, C. 1990. Cometary delivery of organic molecules to the early earth. *Science 249:* 366–373.

Cronin, J. R., and Moore, C. B. 1971. Amino acid analyses of the Murchison, Murray, and Allende carbonaceous chondrites. *Science 172:* 1327–1329.

Decker, P., Schweer, H., and Pohlmann, R. 1982. Identification of formose sugars, presumable prebiotic metabolites, using capillary gas chromatography/gas chromatography-mass spectrometry of *n*-butoxime trifluoroacetates on OV-225. *J. Chromatography 225:* 281–291.

Ferris, J. P., and Chen, C. T. 1975. Chemical evolution. XXVI. Photochemistry of methane, nitrogen, and water mixtures as a model for the atmosphere of the primitive earth. *J. Am. Chem. Soc. 97:* 2962–2967.

Ferris, J. P., Joshi, P. D., Edelson, E. H., and Lawless, J. G. 1978. HCN: A plausible source of purines, pyrimidines and amino acids on the primitive earth. *J. Mol. Evol. 11:* 293–311.

Ferris, J. P., Sanchez, R. A., and Orgel, L. E. 1968. Studies in prebiotic synthesis. III. Synthesis of pyrimidines from cyanoacetylene and cyanate. *J. Mol. Biol. 33: 693–704.*

Ferris, J. P., Zamek, O. S., Altbuch, A. M., and Frieman, H. 1974. Chemical evolution. XVIII. Synthesis of pyrimidines from guanidine and cyanoacetaldehyde. *J. Mol. Evol. 3:* 301–309.

Fox, S. W., and Dose, K. 1972. *Molecular Evolution and the Origin of Life* (New York: Marcel Dekker), 370 pp.

Fuller, W. D., Sanchez, R. A., and Orgel, L. E. 1972. Synthesis of purine nucleosides. *J. Mol. Biol. 67:* 25–33.

Gabel, N. W., and Ponnamperuma, C. 1967. Model for origin of monosaccharides. *Nature 216:* 453–455.

Gilbert, W. 1986. The RNA world. *Nature 319:* 618.

Groth, W., and von Weyssenhoff, H. 1960. Photochemical formation of organic compounds from mixtures of simple gases. *Planet. Space Sci. 2:* 79–85.

Haldane, J. B. S. 1929. The origin of life. *Rationalist Annual 148:* 3–10.

Handschuh, G. J., and Orgel, L. E. 1972. Struvite and prebiotic phosphorylation. *Science 179:* 483–484.

Horowitz, N. H. 1945. On the evolution of biochemical synthesis. *Proc. Natl. Acad. Sci. USA 31:* 153–157.

Hulshof, J., and Ponnamperuma, C. 1976. Prebiotic condensation reactions in an aqueous medium. A review of condensing agents. *Origins of Life 7:* 197–224.

Joyce, G. F., Schwartz, A. W., Miller, S. L., and Orgel, L. E. 1987. The case for an ancestral genetic system involving simple analogues of the nucleotides. *Proc. Natl. Acad. Sci. USA 84:* 4398–4402.

Kenyon, D. H., and Steinman, G. 1969. *Biochemical Predestination* (New York:McGraw-Hill), 301 pp.

Khare, B. N., and Sagan, C. 1971. Synthesis of cystine in simulated primitive conditions. *Nature 232:* 577–578.

Kvenvolden, K. A., Lawless, J. G., Pering, K., Peterson, E., Flores, J., Ponnamperuma, C., Kaplan, I. R., and Moore, C. 1970. Evidence for extraterrestrial amino-acids and hydrocarbons in the Murchison meteorite. *Nature 228:* 923–926.

Kvenvolden, K. A., Lawless, J. G., and Ponnamperuma, C. 1971. Nonprotein amino acids in the Murchison meteorite. *Proc. Natl. Acad. Sci. USA 68:* 486–490.

Lawless, J. G., and Boynton, C. G. 1973. Thermal synthesis of amino acids from a simulated primitive atmosphere. *Nature 243:* 405–407.

Lawless, J. G., Zeitman, B., Pereira, W. E., Summons, R. E., and Duffield, A. M. 1974. Dicarboxylic acids in the Murchison meteorite. *Nature 251:* 40–42.

Lohrmann, R., Bridson, P. K., and Orgel, L. E. 1980. Efficient metal-ion catalyzed template-directed oligonucleotide synthesis. *Science 208:* 1464–1465.

Lohrmann, R., and Orgel, L. E. 1973. Prebiotic activation processes. *Nature 244:* 418–420.

Lohrmann, R., and Orgel, L. E. 1976. Template-directed synthesis of high molecular weight polynucleotide analogues. *Nature 261:* 342–344.

Maher, K. A., and Stevenson, D. J. 1988. Impact frustration of the origin of life. *Nature 331:* 612–614.

Mann, A. P. C., and Williams, D. A. 1980. A list of interstellar molecules. *Nature 283:* 721–725.

Miller, S. L. 1953. Production of amino acids under possible primitive earth conditions. *Science 117:* 528–529.

Miller, S. L. 1955. Production of some organic compounds under possible primitive earth conditions. *J. Am. Chem. Soc. 77:* 2351–2361.

Miller, S. L. 1957. The formation of organic compounds on the primitive earth. *Ann. N.Y. Acad. Sci. 69:* 260–274 [Also in: 1959. A. Oparin (Ed.), *The Origin of Life on the Earth* (Oxford: Pergamon Press), pp. 123–135].

Miller, S. L. 1987. Which organic compounds could have occurred on the prebiotic earth? *Cold Spring Harbor Symp. Quant. Biol. 52:* 17–27.

Miller, S. L., and Orgel, L. E. 1974. *The Origins of Life on the Earth* (Englewood Cliffs, NJ: Prentice-Hall), 229 pp.

Miller, S. L., Urey, H. C., and Oro, J. 1976. Origin of organic compounds on the primitive earth and in meteorites. *J. Mol. Evol. 9:* 59–72.

Miller, S. L., and Van Trump, J. E. 1981. The Strecker synthesis in the primitive ocean. In Y. Wolman (Ed.), *Origin of Life* (Holland: Reidel, Dordrecht), pp. 135–141.

Oparin, A. I. 1938. *The Origin of Life* (New York: Macmillan), 270 pp.

Oro, J., and Kimball, A. P. 1961. Synthesis of purines under primitive Earth conditions. I. Adenine from hydrogen cyanide. *Arch. Biochem. Biophys. 94:* 221–227.

Oro, J., and Kimball, A. P. 1962. Synthesis of purines under possible primitive earth conditions. II. Purine intermediates from hydrogen cyanide. *Arch. Biochem. Biophys. 96:* 293–313.

Oro, J., Nakaparksin, S., Lichtenstein, H., and Gil-Av, E. 1971. Configuration of amino acids in carbonaceous chondrites and a Precambrian chert. *Nature 230:* 107–108.

Peltzer, E. T., and Bada, J. L. 1978. α-Hydroxy carboxylic acids in the Murchison meteorite. *Nature 272:* 443–444.

Peltzer, E. T., Bada, J. L., Schlesinger, G., and Miller, S. L. 1984. The chemical conditions on the parent body of the Murchison meteorite: Some conclusions based on amino, hydroxy and dicarboxylic acids. *Adv. Space Res. 4* (12): 69–74.

Ponnamperuma, C., and Peterson, E. 1965. Peptide synthesis from amino acids in aqueous solution. *Science 147:* 1572–1574.

Reid, C., and Orgel, L. E. 1967. Synthesis of sugar in potentially prebiotic conditions. *Nature 216:* 455.

Ring, D., Wolman, Y., Friedman, N., and Miller, S. L. 1972. Prebiotic synthesis of hydrophobic and protein amino acids. *Proc. Natl. Acad. Sci. USA 69:* 765–768.

Rohlfing, D. L. 1976. Thermal polyamino acids: Synthesis at less than 100°C. *Science 193:* 68–70.

Sagan, C., and Khare, B. N. 1971. Long-wavelength ultraviolet photoproduction of amino acids on the primitive earth. *Science 173:* 417–420.

Sanchez, R. A., Ferris, J. P., and Orgel, L. E. 1966. Cyanoacetylene in prebiotic synthesis. *Science 154:* 784–786.

Sanchez, R. A., Ferris, J. P., and Orgel, L. E. 1967. Studies in prebiotic synthesis. II. Synthesis of purine precursors and amino acids from aqueous hydrogen cyanide. *J. Mol. Biol. 30:* 223–253.

Sanchez, R. A., Ferris, J. P., and Orgel, L. E. 1968. Studies in prebiotic synthesis. IV. The conversion of 4-aminoimidazole-5-carbonitrile derivatives to purines. *J. Mol. Biol. 38:* 121–128.

Schlesinger, G., and Miller, S. L. 1983a. Prebiotic synthesis in atmospheres containing CH_4, CO, and CO_2. I. Amino acids. *J. Mol. Evol. 19:* 376–382.

Schlesinger, G., and Miller, S. L. 1983b. Prebiotic synthesis in atmospheres containing CH_4, CO, and CO_2. II. Hydrogen cyanide, formaldehyde and ammonia. *J. Mol. Evol. 19:* 383–390.

Schwartz, A. W., and Chittenden, G. J. F. 1977. Synthesis of uracil and thymine under simulated prebiotic conditions. *Biosystems 9:* 87–92.

Shapiro, R. 1988. Prebiotic ribose synthesis: A critical analysis. *Origins of Life Evol. Biosphere 18:* 71–85.

Stribling, R., and Miller, S. L. 1987. Energy yields for hydrogen cyanide and formaldehyde syntheses: The HCN and amino acid concentrations in the primitive ocean. *Origins of Life Evol. Biosphere 17:* 261–273.

Urey, H. C. 1952. The Planets (New Haven, Conn.: Yale Univ. Press), 291 pp.

Weber, A. L., and Orgel, L. E. 1978. Amino acid activation with adenosine 5'-phosphorimidazolide. *J. Mol. Evol. 11:* 9–16.

Wolman, Y., Haverland, W. J., and Miller, S. L. 1972. Nonprotein amino acids from spark discharges and their comparison with the Murchison meteorite amino acids. *Proc. Natl. Acad. Sci. USA 69:* 809–811.

Zahnle, K. J. 1986. The photochemistry of hydrocyanic acid (HCN) in the Earth's early atmosphere. *J. Geophys. Res. 91:* 2819–2834.

Zeitman, B., Chang, S., and Lawless, J. N. 1974. Dicarboxylic acids from electric discharge. *Nature 251:* 42–43.

Zhao, M., and Bada, J. L. 1989. Extraterrestrial amino acids in Cretaceous/Tertiary boundary sediments at Stevns Klint, Denmark. *Nature 339:* 463–465.

CHAPTER 2

THE OLDEST FOSSILS
AND WHAT THEY MEAN

■

J. William Schopf†

■

INTRODUCTION

Geologic time, encompassing the total history of the earth, is divided into two eons of markedly unequal duration. The earlier, longer **Precambrian Eon** extends from the time of formation of the planet, some 4550 million years (Ma) ago, to the first appearance in the fossil record of hard-shelled invertebrate animals, about 550 Ma ago. The later, shorter **Phanerozoic Eon** spans the most recent 550 Ma and the familiar evolutionary progression from seaweeds to flowering plants, from invertebrates to humans.

These two vast eons encompass decidedly differing stages in the development of life and of the planet. The Precambrian, comprising the earlier seven-eighths of geologic time, was "The Age of Microscopic Life," an eon dominated by primitive, slowly evolving microbes. Notably, and somewhat surprisingly, virtually the entire basic organization of the present-day biologic world dates from this remote Precambrian past. For example, the basis of the modern food chain—the subdivision of life into plantlike (autotrophic) "producers" and animal-like (heterotrophic) "consumers"—originated during the Precambrian. Similarly, the fundamental partition of the living world into organisms that absolutely require molecular oxygen, and those that can thrive without it, was an ancient, Precambrian innovation. Indeed, numerous biologic inventions of lasting influence (photosynthesis, the ability to breathe oxygen, various types of cell division, sexual reproduction, and many others) first became established among early evolving, "simple" Precambrian micro-organisms.

In contrast, the Phanerozoic Eon—a much shorter segment of earth history, encompassing only the last one-eighth of geologic time—was, as its name implies, "The Age of Megascopic Life" (*phaneros,* Greek, visible, evident; *zoe,* Greek, life). Because the Phanerozoic was far shorter than the preceding Precambrian, we might be tempted to guess

†Department of Earth and Space Sciences, Molecular Biology Institute, and IGPP Center for the Study of Evolution and the Origin of Life, University of California at Los Angeles, Los Angeles, CA 90024 USA

that it was also far less important in the overall evolutionary story. From a global, ecologic perspective, perhaps this is so. But, from a human perspective, just the opposite seems true. After all, we are most interested in our more immediate roots—where did humans come from—and those roots seem to lie in the large organisms of the Phanerozoic. Humans are bilaterally symmetrical, mammalian, self-conscious. But why are we bilateral? Why mammalian? Why are we self-aware? We cannot expect primitive Precambrian microbes to hold the answers. Moreover, as we address such questions from a human perspective, we are saddled with the limitations of our own sensory apparatus. Seeing is believing. What we see, we know, we accept. And what is beyond our ability to perceive, what is too small to be seen, goes unnoticed and unknown.

Appreciation of the influence of these human traits, of the humanness of the situation, contributes much toward explaining why the history of Phanerozoic life has been actively studied since the early 1800s, whereas, until quite recently, the Precambrian fossil record remained shrouded in mystery.

In comparison with the life forms of the Phanerozoic, those of the Precambrian seem vastly different. Most Precambrian organisms are microscopic, virtually too small to be seen! Moreover, these ancient, minute microbes are so distinctly unlike humans that it is difficult to imagine that they could be relevant to understanding the nature of our species. But might this view be mistaken? Might the early, Precambrian evolutionary progression tell us something about ourselves after all? Why, for example, do we breathe oxygen? Why are we dependent on plants, both for the oxygen we breathe and, ultimately, for the food we eat? Why is each of us similar to, but not identical with, our parents or our sisters or our brothers? And, more broadly, why is it that the biology of this planet operates the way it does? How far back into the geologic past can that biology be traced? How has that biology changed over time?

The answers to these questions lie in an understanding of the earlier seven-eighths of the evolutionary story. This earlier, Precambrian history of life is the subject of the following discussion, a history that has only recently come into focus.

Emergence of a New Field of Science

The roots of Precambrian paleobiology extend to the writings of Charles Darwin and to the publication, in 1859, of *The Origin of Species* in which Darwin pointed out, in assessing problems confronted by his theory of evolution, that

> There is another . . . difficulty, which is much more serious. I allude to the manner in which species belonging to several of the main divisions of the animal kingdom suddenly appear in the lowest known [Cambrian-age] fossiliferous rocks. . . . If the theory [of evolution] be true, it is indisputable that before the lowest Cambrian stratum was deposited, long periods elapsed . . . and that during these vast periods, the world swarmed with living creatures . . . [However], to the question why we do not find rich fossiliferous deposits belonging to these assumed earliest periods prior to the Cambrian system, I can give no satisfactory answer. The case at present must remain inexplicable; and may be truly urged as a valid argument against the views here entertained.

Thus the problem posed by the "missing pre-Cambrian fossil record" was well recognized in the mid-1800s. But despite this early recognition of the problem, and several potentially fruitful discoveries during the early 1900s, the question was to remain unanswered—the case to remain "inexplicable"—until more than a century had passed.

In the mid-1960s, however, more than 100 years after Darwin's statement of the problem, this dormant area of science finally began to awaken, spurred by a series of independent, but ultimately synergistic, discoveries. Stimulated by S. L. Miller's pioneering studies during the 1950s (see Chapter 1), interest had been rekindled in the age-old problem of the origin of life. Fossil invertebrate animals, possibly of Precambrian age, had been reported from South Australia (see Chapter 3). Soviet work suggested that microscopic "sporelike" fossils could be recovered from Precambrian shales. The first firm evidence of Precambrian fossil microbes had been discovered in ancient rocks of the Canadian Shield. And modern microbe-containing stromatolites, mound-shaped structures like those known from many Precambrian terranes, had been found in a hypersaline lagoon on the western coast of Australia (see Figure 2.3A). Moreover, by the mid-1960s, NASA's space program was in full swing, providing impetus (and funding) for studies of the earliest history of life. Coupled with new techniques that permitted the ages of ancient rocks to be known with ever-increasing precision, each of these developments played a role in setting the stage for major progress in the field. After a century of inactivity, important pieces of the Precambrian paleobiologic puzzle had at last begun to fall in place.

A marked rise of interest and activity in studies of Precambrian life began about 1965. Figure 2.1 shows that as recently as the mid-1960s, fewer than 100 microfossil occurrences were known worldwide from geologic units dating from the later half of the Precambrian (that is, from sediments deposited during the Proterozoic Era, between 2500 and 550 Ma ago). By the late 1980s, the number of recorded occurrences had grown to more than 2800, an increase of nearly 30-fold. Some 85% of these occurrences were first reported during the two decades beginning in 1970. Virtually all other measures of activity in the field show similar striking changes. Comparable plots could be shown for increases during the past 20 years in the number of Precambrian paleobiologists worldwide; the volume of rock investigated; the amount of data generated; or even the amount of space devoted to such studies in college textbooks.

Interestingly, the graph shown in Figure 2.1 illustrates that not only the quantity, but also the quality of work has markedly increased. Because they are so small, accurate identification of Precambrian microfossils is difficult—they can be easily confused with mineral grains or other inorganic microstructures in a rock. Even modern contaminants, such as living microorganisms (for example, microscopic algae growing in laboratory water) or the windblown spores of modern plants inadvertently introduced during sample preparation, have been mistakenly identified as Precambrian microfossils. Problems of this sort were particularly common when the field was young and there was so little known that it was impossible to guess what the real Precambrian fossil record might actually contain. Figure 2.1 shows, for example, that from 1965 to 1970, nearly 20% of objects described as Proterozoic microfossils were in fact mistakenly identified contaminants and mineral artifacts. Nearly half of the scientific papers published during that period contained at least one such error. However, since 1970, as experience accumulated, the field has matured. By

FIGURE 2.1

Cumulative curves showing the number of occurrences of Proterozoic microfossils and "microfossil-like" nonfossils (pseudofossils and contaminants) reported between 1899 and 1989.

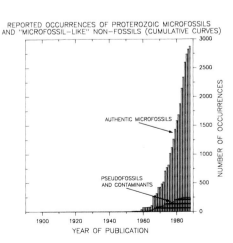

REPORTED OCCURRENCES OF PROTEROZOIC MICROFOSSILS AND "MICROFOSSIL-LIKE" NON-FOSSILS (CUMULATIVE CURVES)

the end of the 1980s, errors had diminished markedly and an active new field of science, Precambrian paleobiology, finally had come of age.

EVIDENCE OF PRECAMBRIAN LIFE

Just as all of geologic time is divided into the Precambrian and Phanerozoic Eons, these eons, in turn, are subdivided into eras. The Precambian Eon is subdivided into two eras: The earlier **Archean Era,** extending from 4550 Ma ago, the time of formation of the planet, to 2500 Ma ago; and the **Proterozoic Era,** spanning the time from the close of the Archean to the beginning of the Phanerozoic, about 550 Ma ago. (Similarly, as is shown in the frontispiece of this book, and discussed in the following chapters, the Phanerozoic Eon is subdivided into three eras, from oldest to youngest, the Paleozoic, Mesozoic, and Cenozoic Eras.) The oldest firm evidence of life now known is about 3500 Ma in age; thus the earliest fossils, and the origin of life, both date from the Archean Era.

The Precambrian Rock Record

Phanerozoic units are relatively abundant, as shown in Figure 2.2, a worldwide inventory of sedimentary rock units of various ages, deposits of the type that might be expected to contain evidence of ancient life. As the inventoried rock units become progressively older, however, they also become decidedly rarer; indeed, relatively few sedimentary rocks older than about 2000 Ma are known, and the rock record older than 3000 Ma is vanishingly small. This correlation between the present-day abundance of sedimentary rocks and their geologic ages is easily explained. The worldwide inventory can contain only those rocks that exist today, those that have survived from the time of their formation to the present. But, as a result of normal geologic processes, rocks are gradually destroyed over time. The effects of weathering, erosion, and the downdrawing into the earth's crust of huge slabs of sediments as a result of global plate tectonics (the movement of massive continental blocks across the earth's surface) combine to obliterate the rock record. As a result of these processes, rock units are destroyed and then geologically reborn, their weathered products ultimately being consolidated to form new, younger sedimentary rock units. In short, the older the rock unit, the greater the chance it no longer exists.

Clearly, the rarity of surviving ancient rocks plays havoc with the search for the earliest records of life. And the search is made even more difficult by other aspects of the available rock record, for the greater the age of any given rock unit, the greater the chance that sometime during its existence it was involved in a mountain-building process and was massively folded and crumpled. During such events, the rocks are pressure-cooked, squeezed, heated, and altered (that is, **metamorphosed**) under conditions that can completely wipe out any evidence of life that the rocks originally contained.

Thus, in the search for the very oldest records of life on earth, there are few surviving, promising places to hunt. Indeed, only two such regions are now known, both containing relatively well-preserved rock units 3500 to 3300 Ma in age. In the discussion of "The Oldest Fossils," later in this chapter, we review recent studies in each of these Archean terranes that have yielded evidence of early life. But first, let us summarize the strategy that has led to these positive results, a strategy based on the search for fossil stromatolites, for preserved fossil microorganisms, and for organic (carbonaceous) matter that can be identified chemically as being a product of ancient life.

FIGURE 2.2

Geologic evidence of changes in the earth's atmosphere and planetary environment over Precambrian and Phanerozoic time.

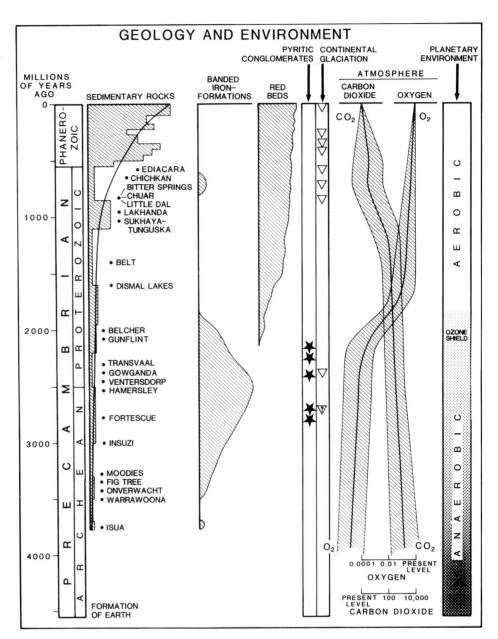

Precambrian Stromatolites and the Nature of Stromatolitic Ecosystems

Columnar or mound-shaped finely layered rock structures known as **stromatolites** have been recognized in the geologic record since the mid-19th century. However, their origin was a matter of dispute; some workers regarded them as nonbiologic concretions, whereas others favored a biological origin. In fact, it was not until the discovery in the early 1960s of living stromatolites (Figure 2.3A) that their biological origin could be definitely shown. They are now known to be biologically produced deposits composed of mineral material that has accreted on and within stacked, matlike layers formed by growing communities of microscopic organisms. Because most stromatolites form in environments in which calcium carbonate ($CaCO_3$) minerals are abundant, fossil stromatolites are common in sedimentary carbonate rocks such as limestones.

FIGURE 2.3

Modern and fossil stromatolites. (A) Living stromatolitic reef composed of solitary and interconnected columnar and mound-shaped carbonate stromatolites at Shark Bay (Hamelin Pool), 650 km north-northwest of Perth, Western Australia; stromatolites are 30 to 40 cm in height. (B) Precambrian stromatolitic reef composed of interconnected columnar and mound-shaped carbonate stromatolites, about 2300 Ma in age, from the Transvaal Supergroup (Campbellrand Subgroup, Transvaal Dolomite) at Groot Boetsap River, 50 km northwest of Warrenton, northern Cape Province, South Africa; stromatolitic reef is about 40 cm in height. (C) Vertical section of a living stromatolitic microbial mat, showing stratified organization of the photic zone—the uppermost growth surface and the immediately underlying undermat layer—and of the thick, underlying oxygen-depleted zone, from "North Pond," Laguna Mormona-Figueroa, 15 km northwest of San Quintin, Baja California del Norte, Mexico.

Modern stromatolites are rare, chiefly because they accrete very slowly and the microorganisms responsible for their formation are eaten by grazing invertebrates, such as snails, before the stromatolites can form. During the Precambrian, however, before the advent of such grazing animals, stromatolitic reefs (Figure 2.3B) were abundant and widespread.

In living examples, the millimeter-thick feltlike mat at the surface of a stromatolite (Figure 2.3C) is almost always formed by an intertwined meshwork of microscopic filamentous cyanobacteria (also known as blue-green algae), but numerous other types of microorganisms occur within such structures. In fact, the diverse organisms occurring within stromatolitic communities form self-contained ecological systems. As shown in Figure 2.3C, these ecosystems typically are organized into three more or less discrete zones:

1. An uppermost matlike layer, the *growth surface,* formed by oxygen-producing photosynthetic cyanobacteria and associated oxygen-requiring (strictly **aerobic**) microbes.

2. A thin immediately underlying layer, the *undermat,* in which occur various types of *non*-oxygen-producing photosynthetic bacteria, as well as **facultative aerobes,** microbes that can use oxygen if it is available, but that also can grow by fermentation in the absence of oxygen.

3. A thick lowermost *oxygen-depleted zone* that contains a veritable menagerie of **anaerobic** microbes, bacteria that thrive in the absence of molecular oxygen.

Because light energy is available in both the growth surface and the undermat, where it is required to power cyanobacterial and bacterial photosynthesis, these two uppermost layers are collectively referred to as the *photic zone* (Figure 2.3C). In shallow waters, where most stromatolites occur, sand grains and other debris tend to wash onto the stromatolitic growth surface, gradually blocking out light and decreasing the light energy available in the photic zone. Were the photosynthetic, light-requiring microorganisms of the zone unable to cope with this slow burial, their energy source ultimately would be cut off; deprived of light, they would die. These microbes avoid this catastrophe, however, by being **phototactic,** that is, by being able to move, to glide, in response to gradients in the intensity of light. As the amount of light available at the growth surface becomes gradually diminished, the phototactic microbes migrate upward through the accumulated debris and establish a new growth surface and a new undermat layer. Depending on local environmental conditions, this process can be repeated, with layer upon layer being added to the slowly accreting stromatolitic structure. Thus, the stacked layers in stromatolites (Figure 2.3C) represent former growth surfaces, laminae produced by the accretion of mineral material on microbial mats formed by communities of actively photosynthesizing microorganisms.

Like all ecological systems, stromatolitic communities are highly ordered, structured by the food relations and the flow of energy within the system. Specifically, they are organized in relation to the availability of light, the local abundance of molecular oxygen, and the nature of the stromatolitic food web.

As we noted, light-requiring stromatolitic microorganisms live in the photic zone, near the top of the community. Interestingly, the photosynthetic bacteria of the undermat layer contain pigments which permit them to absorb light energy in different portions of the spectrum from that used for photosynthesis by cyanobacteria in the immediately overlying growth surface. In essence, the photosynthetic bacteria are able to "see through" the growth surface and thus to carry out their non-oxygen-producing **(anoxic) photosynthesis** (Table 2.1). Because virtually all usable light energy is absorbed in the photic zone, the underlying, thick lowermost zone of the community lacks light-requiring microbes.

As a result of this light-dependent zonation, stromatolitic communities are also stratified with respect to the availability of molecular oxygen. The dominant microbes of the growth surface, cyanobacteria, are oxygen-producing **(oxygenic) photosynthesizers** (Table 2.1); all of the microbes of this uppermost layer are oxygen users (that is, all are able to "breathe," to carry out **aerobic respiration;** Table 2.1); and most cannot survive without oxygen (and are therefore said to be obligately aerobic). Because photosynthesis in the growth surface is dependent on light, oxygen production varies over time. During daylight, oxygen is produced. Although most of this oxygen diffuses upward into the atmosphere or overlying water column, some of it is locally trapped in the mat, where it can seep downward into the undermat layer. At night, in the absence of photosynthesis, oxygen levels decline.

Because the amount of oxygen that diffuses downward from the growth surface varies, both locally and over time, its influence on microbial growth in the undermat also varies. If oxygen is available, it can be used by facultative aerobes of the undermat to carry out aerobic respiration. Because the energy produced via aerobic breakdown of organic compounds is far greater than that yielded by the anaerobic process of **fermentation** (Table 2.1), these microbes are highly efficient oxygen scavengers. Facultative aerobes, therefore, normally keep local oxygen concentrations in the undermat at a low level.

The anoxic photosynthesis carried out by the photosynthetic bacteria of the undermat is a strictly anaerobic process. Hence, like the co-occurring facultative aerobes, these microorganisms respond to local variations in oxygen concentrations. If the amount of oxygen in the immediate environment of the photosynthetic bacteria becomes too high, they "turn off" their photosynthetic apparatus and switch to alternate metabolic processes.

Virtually all of the available oxygen in a stromatolitic community is used either by the aerobes in the growth surface, or by the facultative aerobes in the undermat layer. Oxygen

TABLE 2.1

Principal Biochemical Pathways of Heterotrophic and Autotrophic Metabolism.

HETEROTROPHY

Anaerobic Fermentation (Primitive):

GLUCOSE SUGAR $\xrightarrow{\text{"Glycolysis"}}$ PYRUVATE $\xrightarrow[\text{Molecular Oxygen)}]{\text{(No addition of}}$ ETHYL ALCOHOL + CARBON DIOXIDE + HEAT   +   2 UNITS OF ENERGY

Aerobic Respiration (Advanced):

GLUCOSE SUGAR $\xrightarrow{\text{"Glycolysis"}}$ PYRUVATE $\xrightarrow[\text{"Citric Acid Cycle"}]{\text{(Addition of Molecular Oxygen)}}$ WATER + CARBON DIOXIDE + HEAT   +   36 UNITS OF ENERGY

AUTOTROPHY

Anoxic Bacterial Photosynthesis (Primitive):

CARBON DIOXIDE + HYDROGEN SULFIDE $\xrightarrow[\text{Bacteriochlorophylls}]{\text{(Light Energy)}}$ GLUCOSE SUGAR + SULFUR

Oxygenic Cyanobacterial Photosynthesis (Advanced):

CARBON DIOXIDE + WATER $\xrightarrow[\text{Chlorophyll } a]{\text{(Light Energy)}}$ GLUCOSE SUGAR + OXYGEN

---

levels in the underlying, thick lowermost portion of the stromatolite are therefore maintained at a very low anaerobic level.

Finally, stromatolitic communities are also stratified as a function of food relations. The foundation of any ecologic system is provided by the primary producers, organisms that are able to obtain their carbon from inorganic sources, such as atmospheric or dissolved carbon dioxide, by assimilating it and converting it with the aid of enzymes into insoluble ("fixed") organic compounds. As a result of this carbon fixation, the primary producers add organic matter to their mass—that is, they grow. Both cyanobacteria and photosynthetic bacteria, the predominant primary producers in stromatolitic communities, are **autotrophs** (*autos*, Greek, self; *trephein*, Greek, to nourish; literally, "self-feeders"), organisms that can use carbon dioxide as the immediate and sole source of their cellular carbon (Table 2.1). And because both cyanobacteria and photosynthetic bacteria use light energy to power their autotrophy, both are **photoautotrophs.** Primary producers are the ultimate source of the foodstuffs that are eaten by other members of the community, the **heterotrophic** (*heteros*, Greek, other; literally, "other-feeders") consumers of the ecosystem (Table 2.1).

Because most of the primary producers in stromatolites are photoautotrophs, carbon fixation and primary production are concentrated in the photic zone. This results in a food-related stratification of the community. Primary producers are abundant in the uppermost layers; aerobic heterotrophic consumers are limited to the growth surface and, when oxygen is available, to the undermat; and anaerobic heterotrophic consumers are dominant in the thick, oxygen-depleted lowermost portion of the community.

In sum, modern stromatolitic communities exhibit a characteristic ecologic zonation: anaerobes, predominantly fermenting heterotrophs, at the bottom; facultative microorganisms and various types of photosynthetic bacteria in an intermediate position; and aerobes, including oxygenic photoautotrophs, at the top. Understanding of this organization provides a basis for ecologic investigation of fossil Precambrian stromatolites. Moreover, it is striking that the ecologic zonation of living stromatolitic communities shows essentially the same *spatial* sequence, from bottom to top, as the *evolutionary* sequence inferred to have occurred over time during the early development of the total global ecosystem: Strict anaerobes first; facultative microorganisms second; and obligate aerobes third. Thus, remarkably, the stratification in stromatolitic communities appears to record the overall sequence of events by which increasingly complex layers of biologic organization were added to the world's biota during the earliest history of life!

Precambrian Microfossils

In addition to ancient stromatolites, the Precambrian rock record contains cellularly preserved fossil microorganisms, a second main source of evidence of early life. Two broad categories of such microfossils occur: Nonnucleated, early-evolving bacteria and cyanobacteria, microorganisms collectively referred to as **prokaryotes;** and nucleated, relatively late-evolving micro-algal phytoplankton, the earliest known **eukaryotic** cells.

The principal characteristics of prokaryotes and eukaryotes are summarized in Table 2.2. Of particular paleobiologic usefulness is the fact that prokaryotes, whether unicellular or colonial, have small cells, generally 1 to 10 micrometers (μm) in size; prokaryotes played the dominant roles in the Precambrian drama. In contrast, eukaryotes, represented in the modern world by protists, fungi, plants, and animals, are commonly megascopic organisms that tend to have relatively large cells, 10 to 100 μm in size. With the exception of planktonic micro-algae (microscopic eukaryotes that are grouped with protozoans as protists), eukaryotes were only minor members of the Precambrian cast of characters.

Precambrian microfossils occur principally in two types of sedimentary deposits: In **cherts,** rocks composed of minute interlocking grains of silica, occurring as the mineral quartz (SiO_2), that have been deposited chemically, petrifying microscopic organisms in the place in which they lived (for example, in the layers of stromatolites); and in **shales,** rocks formed by consolidation of layers of clay or mud sedimented, along with phytoplankton and other debris, at the bottoms of lakes or ocean basins.

Although commonly twisted or otherwise distorted, the microfossils of chert deposits are generally unflattened, composed of three-dimensionally preserved organic-walled cells that are thoroughly embedded in, and infilled by, the petrifying fine-grained quartz. In contrast, the carbonaceous microfossils of shales have been preserved by compression, flattened between thin layers of consolidated silt.

Because of the different characteristics of cherts and shales, fossils contained in these two types of rocks are studied by different techniques. Microfossils preserved in cherts are usually studied in very thin slices of the rock, nearly transparent **petrographic thin sections** that, with use of an optical microscope at high magnification, permit the minute petrified microorganisms to be examined in place, entombed within their quartz matrix. Because shales are generally much less translucent than cherts, they are not as easily studied in thin sections. For this reason, the compressed carbonaceous microfossils of shaley deposits are usually examined in **palynological macerations,** concentrated organic-rich residues prepared by dissolving the rocks in mineral acids.

Although use of these techniques has yielded highly significant results, the search for

TABLE 2.2	Comparison of the Principal Characteristics of Prokaryotic and Eukaryotic Organisms.	
	Prokaryotes	**Eukaryotes**
ORGANISMS REPRESENTED:	Bacteria and Cyanobacteria	Protists, Fungi, Plants, Animals
CELL SIZE:	Small, Generally 1 to 10 μm	Large, Generally 10 to 100 μm
CELLULAR ORGANIZATION:	Unicellular or Colonial	Mainly Multicellular with Tissues, Organs
ORGANISM SIZE:	Microscopic	Mainly Megascopic
METABOLISM, PHOTOSYNTHESIS:	Anaerobic, Facultative, Aerobic	Aerobic
FLAGELLA, CILIA:	If Present, Flagella Made of Flagellin Protein	Flagella or Cilia with Microtubules
CELLS WALLS:	Made of Particular Sugars and Peptides	Made of Cellulose or Chitin, But Lacking in Animals
ORGANELLES:	No Membrane-bounded Organelles	Membrane-bounded Chloroplasts and Mitochondria
GENETIC ORGANIZATION:	Loop of DNA in Cytoplasm	DNA in Chromosomes in Membrane-bounded Nucleus
REPRODUCTION:	By Binary Fission; Dominantly Asexual (Some Parasexual)	By Mitosis or Meiosis; Dominantly Sexual

microfossils in Precambrian rocks is often frustratingly unsuccessful. Most Precambrian sediments, like most strata of the Phanerozoic, are simply unfossiliferous, either because fossils were never deposited or because they have been destroyed as a result of normal geologic processes. The search in Precambrian terranes is especially difficult because of the delicate nature and minute size of the fossils, and because many Precambrian sediments have been metamorphically altered by heat and pressure. In general, the most fruitful strategy has been to focus effort on organic-rich ("coal-like," dark brown to black), relatively well-preserved (little metamorphosed) fine-grained shales, and bedded, fine-grained chert layers, especially if the cherts are associated with, or occur within, fossil stromatolites.

Precambrian Organic Matter

All living systems are composed of organic compounds, chemical combinations of the elements carbon (C), hydrogen (H), oxygen (O), nitrogen (N), and sometimes sulfur (S) and phosphorus (P)—"CHONSP." This carbonaceous organic matter is commonly preserved in ancient rocks; coals, for example, are the compressed, preserved remnants of ancient plant debris. Thus it seems reasonable to suppose that the mere presence of carbonaceous matter in Precambrian sediments constitutes firm evidence of the existence of life. In general, this is true (although some of the organic matter in very early Archean sediments might be nonbiological, the preserved products of prebiotic syntheses like those discussed in Chapter 1). Nevertheless, the cardinal question is, "What is the biologic source of this ancient carbonaceous matter—what type(s) of living systems originally made it?"

Fortunately, the biochemical organization of living systems provides the means to answer this question. Organisms are not simply haphazard aggregates of CHONSP-type organic compounds. On the contrary, they are finely tuned organic systems, powered by energy supplied from external sources (for example, sunlight) and organized internally by an ordered, complex network of biochemical reactions. These reactions, in turn, are facilitated, that is, they are catalyzed, by protein-type compounds known as **enzymes,** with each biochemical reaction catalyzed by a specific enzyme. The first biochemical step in both anoxic and oxygenic photosynthesis (Table 2.1), for example, is a reaction resulting in fixation of the carbon atom of carbon dioxide into an organic compound. Both in photosynthetic bacteria and in cyanobacteria, this reaction is catalyzed by an enzyme known as ribulose bisphosphate carboxylase/oxygenase or, in shorthand, **Rubisco.** Because of the unique properties of the Rubisco enzyme, it leaves a telltale isotopic signature in its products, a signature that can be deciphered in organic matter having an age even as great as 3500 Ma.

In nature, carbon atoms exist in three different forms, or **isotopes,** differentiated from each other by their subatomic structure. One of these isotopes, carbon-14 (written in shorthand as ^{14}C), is radioactive. Because it is unstable, disintegrating over time, it cannot be detected in materials older than about 50,000 to 60,000 years. (But because its rate of disintegration is known, it can be used for ^{14}C-dating of very recent organic materials, those from the youngest portion of the Pleistocene Epoch, the most recent phase of the "The Age of Hominids"; see Chapter 6.)

The other two isotopes of carbon, ^{13}C and ^{12}C, are both stable; neither disintegrates over time. Thus, throughout all of earth history, two types of carbon dioxide have existed in the atmosphere, $^{13}CO_2$ and $^{12}CO_2$. The carbon-fixing enzyme in photosynthesis, Rubisco, has the notable property of discriminating between these two types of carbon dioxide, preferentially tending to react with $^{12}CO_2$ and, therefore, to fix preferentially the isotopically lighter stable carbon isotope, ^{12}C, into the organic compounds produced. Thus, both in photosynthetic bacteria and in cyanobacteria (and in all other photoautotrophs as well), the products of Rubisco-catalyzed photosynthesis are enriched somewhat in ^{12}C relative to the concentration of this isotope in the atmospheric CO_2. Like health-food buffs, "photoautotrophs eat light" (both sunlight and isotopically light carbon)!

Atmospheric carbon dioxide is also involved in inorganic chemical processes: CO_2 dissolved in oceanic waters is converted to bicarbonate (HCO_3^-), which can react with calcium to form calcium carbonate, $CaCO_3$, the mineral material of which limestones are composed. As a result of these inorganic processes, and in contrast with the ^{12}C-*enriched* products of photosynthesis, the carbonate carbon in limestones is *depleted* slightly in ^{12}C relative to the atmospheric CO_2 source.

Although the amounts of isotopic discrimination resulting from these various processes are small, they can be measured easily by use of a mass spectrometer. In typical present-day environments, biologically produced organic carbon is *enriched* in ^{12}C by about 17 parts per thousand ($17^o/_{oo}$), relative to atmospheric CO_2, whereas the inorganic, carbonate carbon of limestones is *depleted* in ^{12}C by about $7^o/_{oo}$, a *net difference* between the two types of carbon of $24^o/_{oo}$. The amount of isotopic discrimination can vary depending on environmental conditions, such as the amount of CO_2 in the earth's atmosphere, a concentration that is thought to have changed, probably quite considerably, over the history of the planet. Nevertheless, a net difference of about $20^o/_{oo}$ to more than $40^o/_{oo}$ between the isotopic compositions of biologic and inorganic carbon can be traced far into the geologic past. Indeed, this isotopic signature has been detected in hundreds of Precambrian rock samples, the oldest some 3500 Ma in age. Together with studies of Precambrian stromatolites and microfossils, these biogeochemical data provide important evidence regarding the existence, and nature, of early life.

■

THE OLDEST FOSSILS

Our discussion of the strategy and tools used in the search for evidence of ancient life leads naturally to a series of questions: "Just how far into the geologic past can the fossil record actually be traced? What evidence has survived of the earliest forms of life? And what were these ancient life-forms really like—did they bear any resemblance to organisms living today?"

A short answer: "The oldest evidence of life now known, stromatolites, microfossils, and Rubisco-type carbon isotopic signatures about 3500 Ma in age, indicates that the early biota was composed of prokaryotic microbes, both heterotrophs and autotrophs, and that by this stage in the history of life, oxygen-producing cyanobacteria, evolutionarily advanced microorganisms, may have already appeared on the scene."

The Oldest Known Rocks

A longer answer, which follows, is far more revealing. As we noted, in the search for the very oldest records of life, there are only two promising Archean locales that are known to have survived to the present: One in eastern South Africa, in and near the Kingdom of Swaziland, and the other in the remote "outback" of northwestern Western Australia. In both of these regions, the distribution of the rock strata has been mapped in detail, and the rock units have been formally described and officially named. Like biologic nomenclature (in which similar living species are combined together in a genus; genera are grouped into families; families, into orders; and so forth), geologic nomenclature follows a hierarchical system. By international agreement, the basic first-order members of this hierarchy are distinctive, mappable geologic units known as **formations.** Formations occurring together, forming a stacked sequence of rock units, are collectively referred to as a geologic **group;** and a thick, stacked sequence of groups is termed a **supergroup.** In eastern South Africa, the oldest Archean sequence of rocks has been named the Swaziland Supergroup; as noted in Figure 2.2, this supergroup is composed, from oldest to youngest, of the Onverwacht, Fig Tree, and Moodies Groups (which, in turn, are composed of a total of 12 geologic formations). Similarly, the oldest sequence in Western Australia is the Pilbara Supergroup, composed, in ascending order, of the Warrawoona, Gorge Creek, and Whim Creek Groups (together containing a total of some 20 geologic formations).

Rocks of the Swaziland and Pilbara Supergroups—both composed of geologic formations 3500 to 3300 Ma in age—are *not* the oldest sequences of rocks known on earth. That distinction belongs to the Isua Supracrustal Group of southwestern Greenland, rocks about 3750 Ma in age (Figure 2.2). The Isua rocks, however, have been severely metamorphosed. Although some of the Isua formations have been shown to be of sedimentary origin (providing direct evidence for the occurrence of liquid water at this very early stage in earth history), and certain of these include remnants of sedimentary carbonate deposits (proving that the Isua atmosphere contained CO_2), the Isua rocks have been so severely altered by heat ($\sim$500°C) and high pressure that they cannot reasonably be expected to harbor any clear-cut record of early life. Thus the Swaziland and Pilbara sediments are not the very oldest rocks known, but they *are* the oldest moderately well-preserved geologic units so far discovered, the only truly promising rocks now known in which to search for the earliest evidence of life. Remarkably, both of these very ancient sequences have yielded just such evidence.

The Oldest Known Stromatolites

Shown in Figure 2.4 are two of the oldest stromatolites yet discovered, from the Swaziland (Figure 2.4A) and Pilbara Supergroups (Figure 2.4B). Neither these, nor any of the other stromatolites known from these two Archean sequences has been shown to contain cellularly preserved microorganisms, microfossils that would provide surefire proof that the stromatolites are actually of biological origin. The absence of microfossils is not surprising; because of mineralogical changes that take place while stromatolites are forming, the *vast* majority of fossil stromatolites (well over 99.99%) do not contain preserved cells. However, in the absence of the firm evidence provided by such microfossils, there is always the possibility that these Swaziland and Pilbara structures may not be biologic. For example, their columnar (Figure 2.4A) and mound-shaped (Figure 2.4B) stromatolite-like form might have been produced by slumping and sliding of soft sedimentary material before it became consolidated into solid rock. Nevertheless, the size, shape, and the finely and wavy laminated organization of these structures are certainly very much like those of younger stromatolites, both fossil and modern. In fact, if these structures had been found in younger Precambrian sediments, where both stromatolites and microfossils are abundant and widespread, there would probably be no question about their origin—as the old saying

FIGURE 2.4

Archean stromatolites and filamentous microfossils. (A) Vertical section of a silicified columnar stromatolite, about 3400 Ma in age, from the Swaziland Supergroup (from the lower third of the Fig Tree Group) at "Graywacke Hill," 30 km southeast of Barberton, eastern Transvaal, South Africa. (B) Vertical section of a silicified mound-shaped stromatolite, about 3500 Ma in age, from the Pilbara Supergroup (Warrawoona Group, Towers Formation) near "North Pole Dome," 120 km southeast of Port Hedland, Western Australia. (C–F) Filamentous microfossils, about 3500 Ma in age, in petrographic thin sections of laminated black-and-white chert from the Swaziland Supergroup (Onverwacht Group, Kromberg Formation) in the Komati River Valley, 30 km south of Barberton, eastern Transvaal, South Africa; arrow in (E) points to tubular, sheathlike, hollow filament; because of their meandering, three-dimensional form, composite photomicrographs have been used to illustrate certain of the petrified filaments (in C, D, and F).

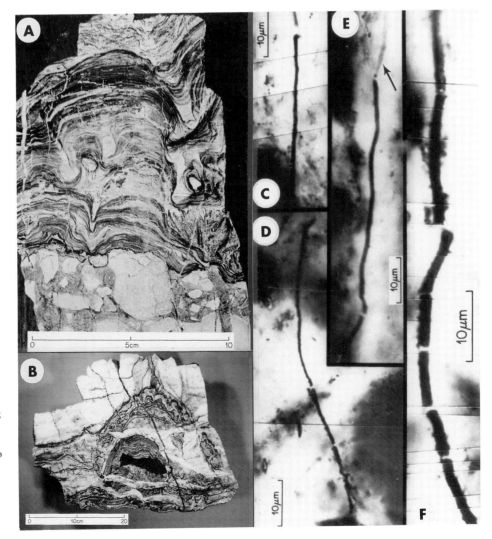

goes, "If it looks like a duck, quacks like a duck, waddles like a duck, well, probably, it *is* a duck!" In the search for the oldest firm evidence of life, however, to merely *seem* to be acceptable evidence may not be quite good enough. We really would like to be able to "prove"—to establish with absolute certainty—that these stromatolite-like structures were produced by once-living communities of ancient microbes. Thus, while the Swaziland and Pilbara stromatolites are highly suggestive of the existence of life, they would probably have to be regarded as less than conclusive by themselves—if they were the only evidence we had.

The Oldest Known Microfossils

The best evidence—evidence that seems fully convincing—comes from microfossils. In particular, narrow, microscopic bacterium- or cyanobacterium-like filaments have been found petrified in sediments of both the Swaziland Supergroup (in finely layered black cherts of the Onverwacht Group; Figure 2.4C–F) and the Pilbara Supergroup (in gray to black carbonaceous cherts of the Warrawoona Group; Figure 2.5A–E).

How can we be sure that these tiny objects are really fossils? To prove the case, three separate tests must be satisfied. First, it must be shown that the fossil-like objects are actually part of the rock, rather than being contaminants. This test has been met: In both deposits, the filaments have been studied in petrographic thin sections; thus it has been demonstrated that the fossils occur *within* the rocks and that they have been present in the rocks since the time when the rocks were formed. Second, the geologic source of the fossiliferous rocks must be known with certainty. This test, also, has been met: The geology of the Swaziland and Pilbara sequences has been studied in detail; in both sequences, fossil-bearing samples have been collected at the fossiliferous localities on repeated occasions; and in both regions, the fossil filaments occur in the lower, and therefore the older, portions of the supergroups—available data indicate that the microfossiliferous rocks are about 3500 to 3400 Ma in age.

To satisfy the third test, it must be shown that the fossil-like objects are definitely the remnants of ancient organisms rather than being, for example, "biologic-shaped" mineral grains. This final test, also, has been met. Isotopic studies indicate that organic matter in both deposits is of biological origin. Moreover, studies by optical microscopy show that in both deposits, the minute petrified filaments, like fossils that have been preserved in a similar manner in younger sediments (petrified woods, for example), are composed of carbonaceous, organic material. Some of the Swaziland filaments appear to be surrounded by a thin, hollow, organic tube (arrow in Figure 2.4E), an organization unlike that occurring in minerals, but one that is well known in modern filamentous bacteria and cyanobacteria (in which strands of cells are encompassed by an originally mucilagelike, tubular, organic "sheath"). And the Pilbara filaments are composed unquestionably of distinct, organic-walled cells (Figure 2.5), arranged in single-file rows like beads on a string—definitely a nonmineralic organization, but one that is characteristic of the great majority of living, filamentous microbes. Furthermore, the particular arrangement of these cells, and the occurrence of partially divided pairs of cells in some of the Pilbara filaments, indicates that they were produced by the same type of cell division as that occurring in living prokaryotic microorganisms.

Six different types of fossil filaments have been discovered in the Pilbara cherts, classified by the size and shape of their cells, including their end cells, which in some filament types are rounded or conical (Figure 2.5A). Interestingly, meticulous comparison of these filaments with microbes living today shows that the majority of the fossils are similar in cellular detail to particular species of living cyanobacteria. Moreover, the occurrence of stromatolites in units both of the Swaziland and Pilbara Supergroups, and the carbon isotopic signature of the organic matter preserved in both of these sequences, are

FIGURE 2.5

Cellularly preserved microbial fossils, about 3500 Ma in age, in petrographic thin sections of bedded, gray to black chert from the Pilbara Supergroup (Warrawoona Group, Apex Basalt), collected in 1982 (A, C–E), and in 1986) (B), along Chinaman Creek, 12 km west of Marble Bar, northwestern Western Australia; because of their meandering, three-dimensional form, composite photomicrographs have been used to illustrate the petrified fossil filaments.

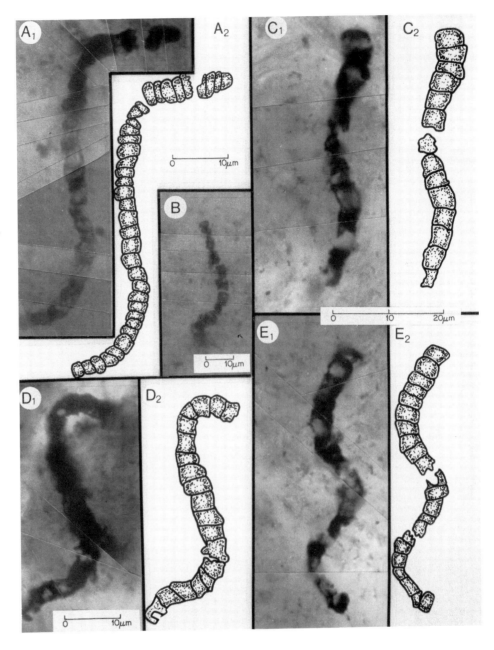

also consistent with the possible presence of cyanobacteria (although they do not prove that cyanobacteria were certainly present, since both the stromatolites and the carbon isotopic values might otherwise have been produced by photosynthetic bacteria). Thus the evidence suggests that cyanobacteria may have existed as early as about 3500 Ma ago. And this is an intriguing possibility, because if these fossil microbes were, in fact, cyanobacteria, they would have to have been capable of oxygen-producing photosynthesis—their presence would indicate that microorganisms with this advanced biochemical capability had already evolved by this early stage in the history of life.

Considering their very great geologic age, the Swaziland and Pilbara filaments, the oldest fossils now known, seem surprisingly advanced. Some are virtually indistinguishable in form from microbes living today and, apparently, their cells divided by the same processes, they lived in the same types of environment, and they carried out the same sorts of metabolism as those of their modern microbial look-alikes. Evidently, shallow-water Archean seas were inhabited by complex, biologically diverse communities of stromatolite-forming microorganisms, self-contained ecosystems that included both photoautotrophic producers and anaerobic, heterotrophic consumers (microbes that recycled the photosynthetically produced foodstuffs), and that possibly even included advanced oxygen-producing cyanobacteria. Clearly, by about 3500 Ma ago, a substantial amount of evolution had already occurred!

■

DEVELOPMENT OF AN OXYGENIC ENVIRONMENT

The Origin of Oxygenic (Cyanobacterial) Photosynthesis: Evidence from Banded Iron-Formations

What is the significance of oxygen-producing photosynthesis—why does it matter? And, if this process might already have evolved by 3500 Ma ago, how much earlier could it have appeared?

Geologic evidence establishes that the starting materials needed for oxygenic photosynthesis (Table 2.1), water and carbon dioxide, were present in the environment very early in earth history. For example, the occurrence of liquid water is evidenced by ripple marks preserved in upper units of the Swaziland Supergroup (specifically, in sediments of the Moodies Group; Figure 2.6A), about 3300 Ma in age. The presence of water is also well evidenced by the occurrence of pillow lavas (pillow-shaped masses of volcanic rock formed when lava flows into a lake or ocean and rapidly cools and solidifies) near the base of the Swaziland Supergroup (in the lower third of the Onverwacht Group; Figure 2.6B) and deposited about 3500 Ma ago. Moreover, as noted earlier, evidence from the highly metamorphosed Isua rocks indicates that both water and carbon dioxide were present in the environment at least as early as 3750 Ma ago. Because the starting materials required for oxygenic photosynthesis were already present at that remote time (and because sunlight, needed to power photosynthesis, has existed since the even earlier time of formation of the solar system), it is worth asking whether oxygen-producing cyanobacteria might also have existed in Isua time. In fact, this is certainly conceivable, a possibility suggested by the occurrence of oxidized iron minerals, occurring in **banded iron-formations (BIFS)** (Figure 2.2), in the Isua sequence.

What are banded iron-formations (abbreviated BIFs), and how do they enter into the picture? In Figure 2.6C, a particularly well-banded BIF is shown in a core sample (a cylindrical rod of rock produced by drilling into the earth's crust, a technique used, for example, in the search for oil). Typically, the banding is produced by an alternation of iron-rich and iron-poor layers in the rock, and because the iron-rich layers are composed of fine rustlike particles of iron oxides (particularly, the mineral hematite, Fe_2O_3; and, in some deposits, the mineral magnetite, Fe_3O_4), the iron-rich layers have a distinctive dull to bright red color. The iron minerals are formed when iron, produced by volcanic activity and dissolved in oceanic waters, combines with molecular oxygen, a chemical reaction that normally occurs in the upper reaches of the water column, where oxygen is present. Because the resulting iron oxides are highly insoluble in seawater, a fine rain of minute rusty particles falls onto the ocean floor.

What is the source of the oxygen needed to form these iron oxides? Unlike nitrogen, water vapor, and carbon dioxide, the other principal gases of the earth's atmosphere, molecular oxygen (O_2) is not released from rocks when they are heated. Thus, although the atmosphere is largely an accumulated product of the volcanic "outgassing" of the earth's

FIGURE 2.6

Geologic evidence of the nature of the Precambrian environment. (A) Evidence of the presence of liquid water early in the Archean: ripple marks, about 3300 Ma in age, from the Swaziland Supergroup (from the lower third of the Moodies Group), at Hollebrand's Pass on Bonanza Gold Mine Road, near Barberton, eastern Transvaal, South Africa. (B) Evidence of the occurrence of volcanism and the presence of liquid water early in the Archean: pillow lava, about 3500 Ma in age, from the Swaziland Supergroup (Onverwacht Group, Komati Formation), in the Komati River Valley, 30 km south of Barberton, eastern Transvaal, South Africa. (C) Evidence of the presence of dissolved oxygen in the early Proterozoic ocean: banded iron-formation, about 2200 Ma in age, in a core sample from the Transvaal Supergroup (Postmasburg Group, Hotazel Formation), 50 km northwest of Kuruman, northern Cape Province, South Africa; core is 2.5 cm in diameter. (D, E) Evidence that the late Archean-early Proterozoic atmosphere did not contain large amounts of free oxygen: (D) pyritic, quartz pebble conglomerate, about 2500 Ma in age, from the Ventersdorp Supergroup (from the "V.C.R.", the Ventersdorp Contact Reef, Venterspost Formation) at Libanon Gold Mine, 25 km west of Johannesburg, Transvaal, South Africa; (E) rounded, detrital pyrite pebbles, shown in cross sections (at arrows), in a cut core sample of the approximately 2500-Ma-old V.C.R. at Libanon Gold Mine.

interior over geologic time, the oxygen in the atmosphere must have a different source. For a great number of reasons, it is well established that the predominant source of atmospheric O_2 is oxygen-producing photosynthesis. From this it follows that prior to the origin of oxygenic, cyanobacterial photosynthesis, the atmosphere must have been virtually devoid of free oxygen; the planetary environment must have been anaerobic (Figure 2.2); and the oxygen required for the deposition of BIFs must have been in very short supply.

Hence, if BIFs were widespread during Isua time, and if these iron oxide-rich units were deposited by the same mechanism as that occurring later in the Precambrian, then their occurrence would seem to require the presence of oxygen-producing photosynthesizers 3750 Ma ago. Unfortunately, however, no one knows whether or not the Isua sediments are typical of rocks deposited worldwide at that time. Outcrops of the Isua rocks are not widespread (most of their areal extent is buried under the vast Greenland ice sheet); those rocks that are exposed have been highly altered by metamorphism (rocks jokingly referred to as being "fubaritic"—"fouled up beyond all recognition"); and there are simply no other well-studied terranes of comparable age now known. At present, therefore, the Isua story is tantalizingly incomplete—perhaps by that time, living systems had originated and had already evolved to the cyanobacterial stage; but perhaps not. The very earliest history of Precambrian life is, as yet, only poorly understood.

So, the possible presence of oxygen-producing photosynthesizers in Isua time (3750 Ma ago) has yet to be resolved; the evidence for Swaziland-Pilbara time (3500 to 3300 Ma ago), while highly suggestive, is not conclusive. When *are* oxygen-producing cyanobacteria known, with certainty, to have been present on the scene?

As is shown in Figure 2.2, BIFs become increasingly abundant in the known rock record between about 3500 Ma ago and the close of the Archean. And it does not much matter where we hunt for BIFs in rocks of about this age—in Australia, India, China, South America, the Soviet Union, Canada, the United States—they occur almost everywhere, reaching a peak about 2500 Ma ago (with the occurrence of the massive BIFs of the Hamersley Group of Western Australia). During this same period, the earliest relatively abundant stromatolites are known to occur, those of the Insuzi Group, of Natal, South Africa, about 3000 Ma in age, and those of the Fortescue Group, of Western Australia, about 2750 Ma in age. Moreover, and although few microfossiliferous deposits are yet known from this period, convincing cyanobacterium-like microfossils have been detected in Fortescue stromatolites. Taken together, these various lines of evidence indicate that oxygenic photosynthesis almost certainly played an important role in the ecosystems of this time. Indeed, given the data available, it is difficult to imagine that cyanobacteria were not both abundant and widespread at least as early as 3000 Ma ago, and probably substantially earlier.

The Anaerobic Early Environment: Evidence from Pyritic Conglomerates

Did the development of oxygen-producing photosynthesis result in an immediate change in the planetary environment from an anaerobic to an aerobic state? In other words, once this enormously important biochemical innovation had become established, did the environment immediately become capable of supporting obligately "air breathing" forms of life (that is, organisms like humans unable to survive without a steady supply of molecular oxygen)? Somewhat surprisingly, perhaps, the answer is no, an answer documented by the minerals and sediments preserved in the early rock record.

Figure 2.2 summarizes the known geologic distribution of **pyritic conglomerates,** a particular type of rock that reveals a good deal about the nature of the early atmosphere. The term *pyritic* means that the rock contains appreciable quantities of the brassy-colored mineral pyrite (iron sulfide, FeS_2), also known as fool's gold; the term *conglomerate* means that the rock is an aggregate of rounded particles, such as the coarse gravel moved and brought together by river currents. One such pyritic conglomerate is shown in Figure 2.6D (from the 2500-Ma-old Ventersdorp Supergroup of South Africa), a coarse-grained sediment deposited at the delta of a Precambrian river, composed of large gray-and-white quartz pebbles between which are lodged bright, shiny grains of pyrite. As shown in Figure 2.6E, the pyrite grains are rounded, rather than being angular; like the larger quartz pebbles of the rock, the pyrite grains were worn down and rounded during their transport to the river mouth. That this sediment was deposited in a river delta is important, because it means that as the grains and pebbles were rounded and transported, they must have come in contact with moving shallow waters, river water that would have been well aerated by the gases of the atmosphere.

Because of their brassy grains and distinctive texture, pyrite-rich conglomerates are easily recognized. If they were abundant today, we would certainly be aware of it. But they are not abundant; indeed, they are virtually unknown. This is because pyrite, produced by the weathering of rocks such as granites, is easily oxidized and then dissolved and destroyed by reaction with molecular oxygen in the aerated rivers of the present day. Thus the occurrence of pyritic conglomerates in the geologic record (many of which also contain well-rounded grains of uraninite, a uranium-rich mineral that is even more easily oxidized than pyrite) indicates that when and where they were deposited, the environment could not have contained large amounts of molecular oxygen. And the fact that such conglomerates occur earlier, but not later, than about 2000 Ma ago (Figure 2.2), provides an overall time frame for atmospheric evolution: Prior to about 2000 Ma ago, the amount of oxygen in the earth's atmosphere must have been kept at a low, more or less anaerobic, level.

Oxygen Sinks

But how could this be so? If oxygen-producing photosynthesis dates from 3000 Ma ago, or even earlier, why was atmospheric oxygen in short supply until a billion years later? Clearly, something must have been scavenging the oxygen, sponging it up and using it before it could build up in the atmosphere. Such sponges are called **oxygen sinks,** and at least three such sinks were available in the early environment.

In addition to nitrogen, water vapor, and carbon dioxide, major constitutents of the atmosphere, volcanoes give off smaller amounts of other gases. Certain of these, such as hydrogen, methane, carbon monoxide, and hydrogen sulfide, can readily combine with molecular oxygen. Thus, for example, hydrogen (H_2) combines with oxygen, that is, it is oxidized, to yield water (H_2O); both methane (CH_4) and carbon monoxide (CO) are oxidized to produce carbon dioxide (CO_2); and hydrogen sulfide (H_2S) is oxidized to yield soluble sulfate (SO_4^{2-}). Such volcanic gasses, therefore, are effective oxygen scavengers, serving as one of the three oxygen sinks in the early environment.

Oxygen produced by photosynthesis was also scavenged biologically. During oxygenic photosynthesis, light energy is used to split water (H_2O) into hydrogen and oxygen; the hydrogen from the water is combined with carbon dioxide to produce organic matter (usually written in shorthand as sugar, "CH_2O"); and the oxygen from the water is released into the environment, and unused by-product of the photosynthetic process. Thus, as shown in Table 2.1, the net chemical reaction of cyanobacterial photosynthesis is

(light energy) + water + carbon dioxide ⟶

organic matter + oxygen.

As is also shown in Table 2.1, however, this is the exact opposite of the biochemical reaction that takes place during aerobic respiration. When organisms "breathe" (a capability of *all* aerobic organisms, whether they are facultative or obligate aerobes, and even including plants, cyanobacteria, and aerobic bacteria), they oxidize organic matter to release energy by carrying out this net chemical reaction:

organic matter + oxygen ⟶

water + carbon dioxide + (cellular energy).

Thus a second oxygen sink was provided by microorganisms capable of carrying out aerobic respiration. The earliest such microbes were no doubt facultative, respiring aerobically when oxygen was available, but switching to anaerobic fermentation (Table 2.1) when local oxygen levels became low (for example, during times of active volcanism, when the oxygen was scavenged by volcanic gases). Facultative aerobes, effective oxygen scavengers, sponged up the photosynthetically produced oxygen before it could build up in the atmosphere.

The third of the three oxygen sinks was provided by the dissolved iron that had accumulated in the world's ocean basins throughout early earth history. As this iron washed up into areas where photosynthetically produced oxygen was available, the oxygen combined with the iron to form insoluble iron oxide minerals; the mineral particles settled to the ocean floor as a fine, rusty rain; and the BIFs were deposited. In effect, the photosynthetically produced oxygen was scavenged from the oceans and buried forever in the form of rust.

The Rise in Atmospheric Oxygen: Establishment of a Stable Aerobic Environment

When and why did this process stop? When did the environment become capable of supporting organisms that could not survive without an uninterrupted supply of molecular oxygen?

Once again, the rock record provides the answer, and the answer is relatively simple. Two of the oxygen sinks have remained essentially unchanged since Archean time: Even in the present day, volcanic gases still scavenge oxygen; and aerobic respiration still continues—as you read this, you are using oxygen produced by photosynthesis. But the third oxygen sponge, the BIF sink, finally became saturated. That is, over time, slowly but irrevocably, all of the iron dissolved in the world's oceans finally became converted into the iron-rich layers of the BIFs, leaving behind massive iron-rich ores that are used today to make steel—the stoves, bridges, steel girders, and automobiles of the present day are a legacy of this ancient time. The earth's oceans had been swept free of dissolved iron; lowly cyanobacteria—pond scum—had rusted the world!

This episode came to a close with deposition of the last of the major BIFs, about 2000 to 1800 Ma ago. Significantly, at about this same time, a second type of iron-containing sediment became increasingly abundant, deposits known as **red beds.** As their name implies, red beds are reddish colored sedimentary rocks, and their color, like that of the BIFs, is imparted by the presence of iron oxide minerals. However, unlike BIFs, which are subaqueous deposits, many red beds are terrigenous sediments (that is, they are deposited on the land surface), and red beds contain only a small amount of iron (usually less than 1% by weight, occurring as thin, rindlike iron oxide coatings on individual grains of sand or silt). In essence, red beds provide evidence of the subaerial oxidation of sands or clays. Their geologic distribution (Figure 2.2) thus indicates that substantial amounts of molecular oxygen must have been present in the earth's atmosphere beginning by about 1800 Ma ago.

OVERVIEW: THE OLDEST FOSSILS AND THE DEVELOPMENT OF AN AEROBIC ENVIRONMENT

The principal points of the preceding two sections can be summarized as follows:

1. The oldest records of life now known, stromatolites, microfossils, and Rubisco-type carbon isotopic values, date from about 3500 Ma ago.

2. By this stage in the evolution of life, complex, microbial stromatolitic ecosystems, possibly including oxygen-producing cyanobacteria, had already become established.

3. Thus the origin of life (see Chapter 1) occurred earlier, and perhaps many hundreds of millions of years earlier, than 3500 Ma ago.

4. By at least 3000 Ma ago, stromatolitic communities, dominated by photosynthetic, oxygen-producing cyanobacteria, were evidently both abundant and widespread.

5. Geologic evidence, in particular the distribution over time of pyritic conglomerates, indicates that oxygen production by cyanobacteria did not result immediately in a major increase in the amount of atmospheric oxygen.

6. Instead, the molecular oxygen was consumed by three oxygen sinks: Reaction with volcanic gases; use by facultatively aerobic microbes; and burial in the iron-rich layers of banded iron-formations (BIFs).

7. Ultimately, by about 2000 to 1800 Ma ago, after the oceans had been slowly, but irrevocably, swept free of dissolved iron, atmospheric oxygen levels began to rise, and a stable aerobic environment—one capable of supporting organisms dependent on a steady, uninterrupted supply of molecular oxygen—finally became established.

■

THE PROTEROZOIC EVOLUTION OF PRIMITIVE PROKARYOTES

As is shown in Figure 2.7, the Archean (pre-2500-Ma-old) fossil record is minuscule; only a few types of microfossils are known, and even fewer fossiliferous units (from only three geologic groups) have as yet been discovered. Similarly, only a handful of fossil-bearing units are known from the earliest several hundred million years of the Proterozoic. In units about 2000 Ma and younger, however, the picture is markedly improved. In fact, from that time to the present, the documented history of life is based on a more or less continuous

FIGURE 2.7

Schematic diagram showing the abundance of known species of prokaryotic and eukaryotic microfossils, and of microfossiliferous rock units, over Precambrian time.

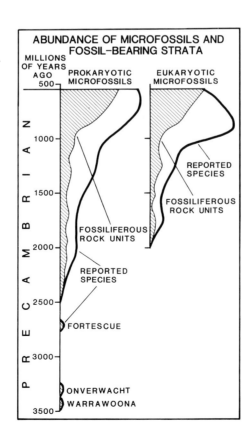

ABUNDANCE OF MICROFOSSILS AND FOSSIL–BEARING STRATA

sequence of fossil-bearing strata. Moreover, the prospects are good that the ancient fossil record will soon become decidedly better known. At present, the number of different types of fossil prokaryotes reported per unit of time during the Proterozoic is roughly parallel to the number of fossiliferous units that have been sampled (Figure 2.7). Hence, although reported prokaryotes become increasingly numerous over the last 500 Ma of the Proterozoic, this is evidently because the number of rock units investigated also increases, rather than because the various types of prokaryotes actually became more numerous. From this it follows that as more rock units are studied, more different types of microfossils are likely to be discovered, and the early history of life should become increasingly better understood.

But, based on the data currently at hand—evidence that is *vastly* improved over that available only a scant two decades ago—what can be said about the Proterozoic evolution of primitive prokaryotes? How did prokaryotes evolve over the nearly two billion years of Proterozoic time?

The answers: A lot can be said; even at present, a great deal is known. Surprisingly, however—in fact, absolutely remarkably—cyanobacteria in particular, and perhaps all prokaryotes in general, seem to have evolved hardly at all between early in the Proterozoic and the present day! That, indeed, is strange. The history of life, at least that of familiar Phanerozoic life, is one of change: Evolution, an unending progression of new forms of life . . . survival of only the fittest . . . the old supplanted by the new. But prokaryotic cyanobacteria seem to have played the evolutionary game by a different set of rules: What once succeeded, continued to succeed . . . survival of the already fit . . . if it's not broken, don't fix it!

To understand this unexpected twist in the evolutionary story, let us dissect the problem into its two components. First, is it really true that cyanobacteria have evolved little, if at all, since Proterozoic time? And second, if that *is* true, how can it be explained—why did these primitive prokaryotes not evolve, whereas eukaryotes obviously did?

FIGURE 2.8

Comparison of living and fossil (Proterozoic) cyanobacteria. Living cyanobacteria (A, C, E, G) are from stromatolitic microbial mats at "North Pond," Laguna Mormona-Figueroa, 15 km northwest of San Quintin, Baja California del Norte, Mexico. (A) *Lyngbya*, a living filamentous (oscillatoriacean) cyanobacterium composed of a single-file row of disk-shaped cells, terminated by rounded cells, that is encompassed by a cylindrical, tubular sheath (at arrow). (B) *Palaeolyngbya*, a *Lyngbya*-like fossil cyanobacterium about 950 Ma in age, shown in a palynological maceration of carbonaceous shale from the Lakhanda Formation of the Khabarovsk region of eastern Siberia, USSR; the arrows point to remnants of the encompassing cylindrical sheath. (C) *Spirulina*, a living filamentous (oscillatoriacean) cyanobacterium having distinctive coiled filaments. (D) *Heliconema*, a *Spirulina*-like fossil cyanobacterium about 850 Ma in age, shown in a palynological maceration of carbonaceous shale from the Miroedikha Formation of the Turukhansk region of eastern Siberia, USSR. (E) *Gloeocapsa*, a living coccoid (chroococcacean) cyanobacterium having a well-ordered, few-celled colony composed of spheroidal to ellipsoidal cells that are enclosed by a thick, distinct sheath (at arrow). (F) *Gloeodiniopsis*, a *Gloeocapsa*-like fossil cyanobacterium about 1550 Ma in age, shown in a petrographic thin section of carbonaceous chert from the Satka Formation of the southern Ural Mountains of Bashkiria, USSR; the arrow points to the preserved, thick enclosing sheath. (G) *Entophysalis*, a living coccoid (entophysalidacean) cyanobacterium having an irregularly ordered, many-celled colony composed of spheroidal to ellipsoidal cells that are embedded in a diffuse, common sheath. (H) *Eoentophysalis*, an *Entophysalis*-like fossil cyanobacterium about 2150 Ma in age, shown in a petrographic thin section of stromatolitic black chert from the Belcher Group (Kasegalik Formation) of the Belcher Islands, Hudson Bay, Northwest Territories, Canada.

The "Arrested Evolution" of Proterozoic Cyanobacteria

The answer to the first of these questions seems clear: Judging from their morphology, living and Proterozoic cyanobacteria are virtually indistinguishable. Figure 2.8 shows just four of a large number of examples that might be cited illustrating this uncanny similarity between the modern and the fossil. Compare, for example, the living cyanobacterium *Lyngbya*, shown in Figure 2.8A, with fossil *Paleolyngbya*, shown in Figure 2.8B: Both are about the same size; both are composed of single-file rows of disk-shaped cells; both have rounded terminal cells; both are enclosed by tubular organic sheaths. The fossil filaments

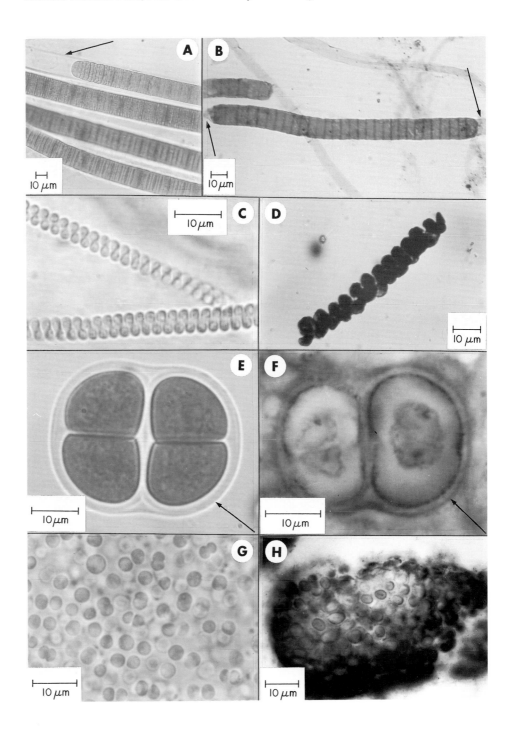

FIGURE 2.9

Comparison of the size range and pattern of size distribution of the tubular sheaths of living (oscillatoriacean) cyanobacteria, with those of the sheaths of fossil prokaryotes.

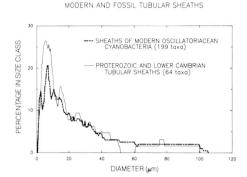

(Figure 2.8B) are nearly one billion years old, but if they were alive today, they surely would be placed in the same genus, and perhaps even the same species, as the modern filaments. Or compare the living coiled cyanobacterium *Spirulina* (Figure 2.8C) with its 850-Ma-old fossil look-alike (Figure 2.8D); or the sheath-enclosed, four-celled colony of modern *Gloeocapsa,* shown in Figure 2.8E, with the notably similar 1550-Ma-old *Gloeocapsa*-like colony shown in Figure 2.8F; or compare the many-celled colony of living *Entophysalis* (Figure 2.8G) with its fossil counterpart, the 2150-Ma-old colonial *Eoentophysalis* (Figure 2.8H). Other evidence is also available. The range of diameter and pattern of size distribution of the tubular sheaths of living cyanobacteria and of comparable fossil prokaryotic filaments are essentially identical (Figure 2.9). Moreover, there is no discernable trend, no evident pattern of evolutionary change, in the size either of fossil tubular sheaths or of fossil cellular filaments over all of Proterozoic time (Figure 2.10), an eon some four times longer than the entire Phanerozoic!

The conclusion seems inescapable—whether the organisms are filamentous or spheroidal, whether they are straight or coiled, whether they occur in few- or in many-celled colonies, and regardless of cell size or cell shape or the nature of their extracellular sheaths—the morphology of cyanobacteria has changed little, if at all, since Proterozoic time.

The Tempo of Cyanobacterial Evolution

In 1944, the American paleontologist George Gaylord Simpson coined three terms to refer to the different tempos of evolution he observed in the Phanerozoic fossil record: **Tachytelic,** or "fast" evolution; **horotelic,** the "standard" rate of evolution; and **bradytelic,** "slow" evolution like that exhibited by lampshells (brachiopods), horseshoe crabs, coelacanth fish, and other so-called living fossils. Most Phanerozoic organisms, such as

FIGURE 2.10

Histograms showing the diameter ranges of cells and of cylindrical tubular sheaths of filamentous prokaryotic microfossils reported from Proterozoic shales and cherts of various ages.

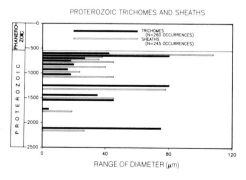

mammals, are horotelic, existing for a few to several million years before they become extinct. In contrast, bradytelic species exhibit little or no morphological change over periods as long as 100 million years or more. But the size, shape, and cellular organization of cyanobacteria have remained unchanged over periods of more than one, and in some cases more than two *billion* years. And Simpson, of course, had no way of knowing what would be found in the 1970s and 1980s in the Precambrian fossil record. Thus a fourth term has been added to Simpson's list: **Hypobradytely** (with the Greek prefix "hypo" meaning low); hypobradytely—the lowest of slow rates of evolution, exhibited by prokaryotic cyanobacteria.

The concept of hypobradytely is not without potential pitfalls. For example, although there are many different types of cyanobacteria, all are morphologically quite simple; if they were structurally more complex, providing more features for comparison between living and fossil species, some portion of their apparent lack of evolution might well disappear. Moreover, hypobradytely, like the other terms used to describe evolutionary rates, is based solely on comparisons of the morphology of modern and fossil organisms, rather than on such important underlying traits as genetics, biochemistry, and physiology. Nevertheless, in addition to morphology, there are many other obvious similarities between living and Proterozoic cyanobacteria (for example, the environments in which they occur; their ability to form stromatolites; their concentration at the growth surface in microbial communities; and the particular types of cyanobacteria, the families and genera, occurring in such communities). Hence, all things considered, it seems quite likely that the fundamental nature of such prokaryotes has evolved little, if at all, since early in Proterozoic time.

The Mode of Cyanobacterial Evolution

If all other organisms evolve at some "reasonable rate," why not cyanobacteria, too? The basic answer is that these primitive prokaryotes simply did not need to evolve—if it's not broken, don't fix it! Moreover, because of their simple asexual means of reproduction, even if cyanobacteria "wanted to evolve" there is good reason to believe that their options would have been limited.

Unlike other, "normal" photoautotrophs—a pine tree, for example, or a clump of grass—cyanobacteria can live almost anywhere: In deserts, tundra, or forest floor cover; beneath mineral crusts, actually *within* rocks, in the Dry Valleys of Antarctica; in hot springs or snowfields, or on the bottoms of permanently ice-covered lakes; in freshwater, marine, or extremely hypersaline (salt-rich) settings; in anaerobic to oxygen-rich environments; and in acidic (pH ~3.5) to exceptionally basic locales (pH 11 to perhaps 13). Although photosynthetic, cyanobacteria can survive prolonged periods of total darkness, some species having been recovered from a depth of 1000 m (far below the level to which sunlight can penetrate) in the Mediterranean Sea and the Indian Ocean. Many are highly resistant to desiccation—several species have been found near Calama, Chile, in the Atacama Desert, a town where rainfall has never been recorded, the driest place on earth. Virtually all are highly resistant to the lethal effects of X rays, UV, and gamma irradiation. In fact, cyanobacteria are extremely difficult to exterminate, even when subjected to the most unusual conditions. One species made the *Guinness Book of World Records* as having been "brought back to life," revived with a bit of tap water, after 107 years of storage as a dried museum specimen. Three other species are reported to have survived immersion in liquid helium at $-269°C$ for 7.5 hours (another world record). And several species have been shown to survive nuclear test-site explosions within a distance of about 1 km from ground zero (and they might even have been able to survive at a closer distance, were it not for the fact that this was the edge of the zone from which all of the soil cover had been stripped away by the atomic blast).

The point is this: Unlike eukaryotes, prokaryotic cyanobacteria are exceptionally gifted ecologic generalists—name the place, they can live there! Moreover, and again unlike virtually all eukaryotes, cyanobacteria are strictly asexual, that is, changes in their genetic material can be provided only via chance mutations rather than by the far more effective process of sexual reproduction.

Taken together, the exceptional survivability of these prokaryotes, their unexcelled ecologic versatility, and the absence of the genetic variability provided in eukaryotes by sexual reproduction, provide a convincing explanation for cyanobacterial hypobradytely— these hardy asexual generalists do not "need" to evolve more rapidly, and they are limited by their genetics in their potential to do so.

THE PROTEROZOIC EVOLUTION OF EUKARYOTIC PHYTOPLANKTON

The last major development in the history of Precambrian microscopic life was the origin of the nucleated eukaryotic cell, the type of cell that makes up all multicellular organisms, including humans (Table 2.2). There seems ample evidence that the earliest eukaryotic cell was a sort of aggregate microorganism, assembled from a consortium of previously free-living prokaryotes. For example, the **mitochondrion,** the intracellular body (organelle) that enables eukaryotic cells to carry out aerobic respiration, is the evolutionary descendant of an originally free-living aerobic bacterium that became incorporated in a host cell early in the evolution of the eukaryotic lineage. Similarly, the **chloroplast,** the intracellular site of photosynthesis in eukaryotic phytoplankton and in multicellular algae and higher plants, is derived from a free-living cyanobacterial ancestor. Thus, in this sense, the eukaryotic cell is the evolutionary derivative of several, more primitive, prokaryotic ancestors.

The time of origin of the eukaryotic cell presents a vexing problem to paleobiologists. There can be no doubt that the earliest eukaryotes were simple microscopic cells, similar in size and shape to certain of the prokaryotes living in their environment. Of course, if these earliest forms were "complete eukaryotes," they would have had a nucleus, mitochondria, and chloroplasts, and because of these intracellular organelles, they would have been easily distinguishable from coexisting prokaryotic microbes. Unfortunately, however, such organelles are very rarely preserved in fossils. Hence, the paleobiologist must rely on other criteria, such as cell size, to differentiate fossils of early eukaryotes from those of prokaryotes. Early eukaryotes can therefore be detected in the fossil record only if they have evolved sufficiently to have become distinguishable from their prokaryotic ancestors. For this reason, rather than being able to pinpoint the time of origin of the eukaryotic cell type, studies of the fossil record can only place a minimum age on the appearance of eukaryotes.

As outlined earlier (in the discussion of the development of an oxygenic environment), a stable, aerobic global environment first became established about 2000 to 1800 Ma ago. Because all complete eukaryotes are strict aerobes, it seems likely that such eukaryotes could not have become widespread before that time. And, perhaps significantly, it is about this time that the first identifiable evidence of eukaryotes appears in the fossil record.

Earliest Evidence of the Eukaryotic Cell

As is shown at the bottom of Figure 2.11, living spheroidal cyanobacteria, the largest coccoid prokaryotes, are generally less than 10 μm in diameter, although a very few species are as large as 60 μm. In contrast, living eukaryotic micro-algae tend to have much larger cells. Thus, judging from the modern biota, fossil coccoid cells in the 10 to 60 μm size

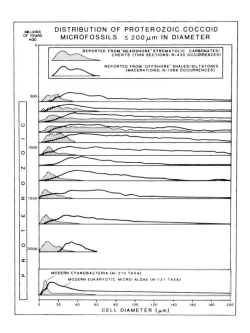

FIGURE 2.11

Comparison of the size ranges and patterns of size distribution of Proterozoic coccoid microfossils ≤200 μm in diameter, reported from "nearshore" stromatolitic carbonates and from "offshore" carbonaceous shales of various ages, with those of modern coccoid cyanobacteria and eukaryotic micro-algae (shown in the rectangle at the bottom of the figure).

range would seem reasonably regarded as "possible eukaryotes," and those larger than 60 μm, larger than the largest living spheroidal prokaryotes, as "assured eukaryotes."

As is also shown in Figure 2.11, the size range of fossil coccoid cells detected in Proterozoic stromatolitic units, sediments deposited in relatively shallow water and, thus, in more or less "nearshore" environments, is decidedly different from that in Proterozoic shales and siltstones, units that tend to be prevalent in relatively "offshore" settings. With but few exceptions, the cells reported from the "nearshore" stromatolites are of cyanobacterial (prokaryotic) size. Thus, like the coccoid microorganisms in modern stromatolites, the vast majority of those preserved in Proterozoic stromatolitic deposits appear to be prokaryotes. In contrast, the "offshore" shales contain much larger cells; those about 2000 Ma in age include possible eukaryotes; those about 1750 Ma and younger include cells within the "assured eukaryotic" size range. These simple, spheroidal fossil micro-algae constitute the oldest evidence yet detected of the eukaryotic cell type.

Proterozoic Acritarchs

Diverse types of planktonic, eukaryotic micro-algae are now known from the Proterozoic. The exact biological relationships of these micro-algae, however, are uncertain (it is unknown, for example, whether they are green algae, red algae, or planktonic representatives of some other group of algal protist). For this reason, they are classified as **acritarchs** (*akritos*, Greek., undecided, confused). Two principal types of acritarchs are known: Simple, spheroidal microfossils (Figure 2.12A, 2.12B) termed **sphaeromorphs** (or, if they are larger than 200 μm in diameter, **megasphaeromorph acritarchs**), and somewhat more complex, spiny spheroids known as **acanthomorph acritarchs** (Figure 2.12D). As is shown in Figure 2.12C, ellipsoidal acritarchs also occur in Proterozoic sediments, as do numerous other varieties.

Relatively small sphaeromorph acritarchs, those 60 to 200 μm in size, are abundant both in Proterozoic- (Figure 2.11) and in Phanerozoic-age strata. Interestingly, however, truly large sphaeromorphs, those more than 1000 μm (one millimeter) in diameter, are typical only of the Proterozoic. As is shown in Figure 2.13, a very few occurrences of such

FIGURE 2.12

Fossil, eukaryotic planktonic micro-algae ("acritarchs"), shown in palynological macerations of Proterozoic carbonaceous shales. (A) A thick-walled, spheroidal micro-alga (the sphaeromorph acritarch, *Valeria*), about 800 Ma in age, from the Akberdin Formation of the southern Ural Mountains of Bashkiria, USSR. (B) A thin-walled, compressed, and highly folded spheroidal micro-alga (the sphaeromorph acritarch, *Kildinella*), about 950 Ma in age, from the Lakhanda Formation of the Khabarovsk region of eastern Siberia, USSR. (C) An ellipsoidal, planktonic micro-alga (*Macroptycha*), about 850 Ma in age, from the Miroedikha Formation of the Turukhansk region of eastern Siberia, USSR. (D) A spiny, planktonic micro-alga (the acanthomorph acritarch, *Trachyhystrichosphaera*), about 950 Ma in age, from the Lakhanda Formation of the Khabarovsk region of eastern Siberia, USSR; arrows point to short, tapered spines.

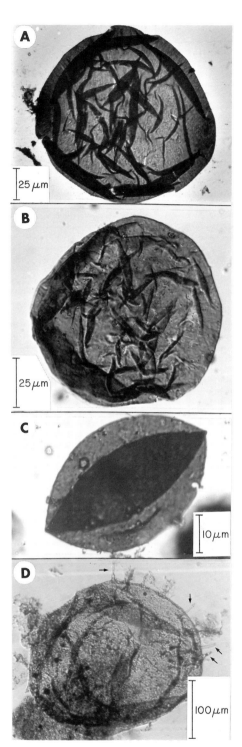

megasphaeromorphs have been reported from sediments older than about 1250 Ma. However, their well-established fossil record begins about 1100 Ma ago and continues up to the beginning of the Phanerozoic. During this period, both their abundance and their diversity waxed and waned—prior to about 1000 Ma ago, megasphaeromorphs were relatively rare, and all were smaller than 3 mm; by 850 Ma ago, the group had become

FIGURE 2.13

Size distributions of eukaryotic, megasphaeromorph acritarchs (spheroidal, planktonic fossil micro-algae >200 μm in diameter) reported from Proterozoic geologic units of various ages.

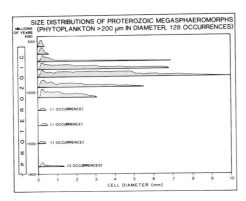

maximally abundant and highly diverse, including exceptionally large species, some nearly a centimeter across; and by 675 Ma ago, the group had become far less abundant, all of the large forms had become extinct, and only a few small-celled species, those less than a few hundred micrometers in diameter, survived into the Phanerozoic.

The details of the late Proterozoic demise of this previously highly successful group are summarized graphically in Figure 2.14. Moreover, as shown in Figure 2.15, acanthomorph acritarchs evidently suffered the same fate during this same period: Large acanthomorphs, hundreds of micrometers in diameter, are known from sediments 950 to 800 Ma in age; but, in deposits of the Lower Cambrian, the oldest sediments of the Phanerozoic Eon, the largest spiny acritarchs known are only about 75 μm in diameter.

Evolution of Proterozoic Eukaryotic Phytoplankton

What happened during the Proterozoic evolution of eukaryotic phytoplankton? Why the wax? Why the wane?

First, the evidence suggests that the eukaryotic cell type appeared early in the Proterozoic, perhaps more or less coincident with the development of a stable aerobic environment some 2000 to 1800 Ma ago.

Second, eukaryotic micro-algae existed, but evolved only quite slowly, prior to about 1200 Ma ago. Like slowly evolving Proterozoic cyanobacteria, these early evolving eukaryotes evidently lacked the capacity for sexual reproduction. Instead, like a few species of micro-algae living today, they probably reproduced solely via body cell division **(mitosis)**, a process yielding offspring that are exact copies of their parent cell.

Third, about 1100 Ma ago, eukaryotic plankton began to diversify very rapidly. Interestingly, it is at about this time that fossil phytoplankton first show plausible evidence

FIGURE 2.14

Maximum diameters of eukaryotic, megasphaeromorph acritarchs reported from geologic units ranging from 1000 to 530 Ma in age; data for spiny phytoplankton (acanthomorph acritarchs) are those shown in Figure 2.15 at an expanded scale.

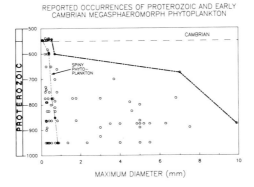

FIGURE 2.15

Maximum diameters of eukaryotic, spiny phytoplankton (acanthomorph acritarchs) reported from geologic units ranging from 1000 to 530 Ma in age.

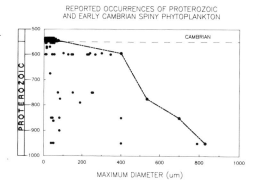

of eukaryotic sexuality, the occurrence of cystlike structures that may have harbored reproductive bodies. If these cysts were part of a sexual life cycle, they would provide evidence both of the occurrence of sex cell division (**meiosis,** an evolutionary derivative of mitotic body cell division), and of the fusion of sex cells (**syngamy**), required to complete the life cycle. The capacity for sexual reproduction was one of the most important of all innovations to have ever evolved, a development providing means for the production of offspring that combine traits from two parental sources, rather than being exact copies of a single, asexual parent cell. Thus the advent of eukaryotic sexuality would have resulted in tremendous increases of genetic diversity and of rates of evolution in the newly evolved sexual lineage. At present, therefore, the rapid rise of planktonic eukaryotes beginning about 1100 Ma ago—the "wax" of the waxing and waning in Proterozoic eukaryotic evolution—seems reasonably explained as a result of the origin and establishment of sexual reproduction.

Fourth, about 950 to 850 Ma ago, Proterozoic eukaryotic acritarchs reached their zenith of abundance and diversity; the phytoplanktonic flora was dominated by large megasphaeromorphs, some several millimeters in diameter, forms that by the end of the Proterozoic were destined to become extinct.

And fifth, between about 900 Ma and perhaps 675 Ma ago, the world's planktonic micro-algal flora experienced a major collapse—the "wane" in Proterozoic eukaryotic history. Although the occurrence of such a decline is well documented (Figure 2.7), the reasons for this collapse, and even its exact timing, are somewhat uncertain. These are new questions, being asked by paleobiologists for the first time, and the data are not all in. As shown in Figure 2.2, however, during this period there appears to have been a decrease in global atmospheric carbon dioxide, a decline possibly responsible for the onset of widespread continental glaciation (as a result, in essence, of a "reversed greenhouse effect"). Moreover, this decrease in atmospheric CO_2 appears to have been coupled with an increase in the concentration of atmospheric oxygen (brought on by rapid burial at this time of photosynthetically produced organic matter). As it turns out, these two gases, CO_2 and O_2, are both able to react with Rubisco, the carboxylase/oxygenase enzyme that drives photosynthesis. Experiments with populations of living micro-algae show that if the concentration of CO_2 is decreased while that of O_2 is increased, photosynthetic activity rapidly diminishes, ultimately reaching a point at which the micro-algae completely cease to grow. Perhaps this is what occurred, on a global scale, during the late Proterozoic.

OVERVIEW: THE PROTEROZOIC EVOLUTION OF PROKARYOTES AND EUKARYOTES

The principal points from the preceding two sections can be summarized as follows:

1. As now known, a "good fossil record," a more or less unbroken continuum of fossil-bearing strata, exists from about 2000 Ma ago to the present.

2. Prokaryotic cyanobacteria, well represented in the Proterozoic fossil record, are now known to have exhibited virtually no evolutionary change in morphology over hundreds, and even thousands, of millions of years.

3. The hypobradytelic evolution of these prokaryotes is a result of their exceptional survivability, their unexcelled ecological flexibility, and their lack of sexual reproduction.

4. The earliest eukaryotes identifiable in the fossil record date from about 1750 Ma ago, more or less coincident with the time of establishment of a stable aerobic environment.

5. Between 1750 and 1200 Ma ago, eukaryotic phytoplankton (acritarchs) evolved quite slowly, probably because they were entirely asexual, capable of reproducing only via body cell division (mitosis).

6. Beginning about 1100 Ma ago, planktonic eukaryotes diversified rapidly, possibly as a result of the advent and establishment of eukaryotic sexuality (based on meiosis and syngamy), to reach their zenith of abundance and diversity about 950 to 850 Ma ago.

7. Late in the Proterozoic, between about 900 and 675 Ma ago, eukaryotic phytoplankton declined markedly in both abundance and diversity, a result, possibly, of a global decrease in atmospheric carbon dioxide and a global increase in atmospheric oxygen.

THE DIFFERENCE BETWEEN "KNOWING" AND "GUESSING"

Clearly, much has been learned over the past two decades about the antiquity and evolutionary history of Precambrian life. Just as obviously, however, a great deal remains unknown. Each new major discovery has led to a myriad of brand-new questions. We now know that life existed at least as early as 3500 Ma ago. But how much earlier did the origin of life occur? And how far had evolution progressed when the oldest known fossils were once alive? We now know that cyanobacteria appeared early, became established, and then evolved little over an enormous span of earth history. But is this typical of prokaryotes in general? And how can we be sure that the nearly identical architectures of modern and fossil microbes do not mask major differences, for example, in their biochemistry and physiology? We now know that eukaryotic evolution started slowly, then rapidly picked up steam, only to meet an impasse late in the Proterozoic. We think we have an explanation for all this—asex, sex, and a reversal of the greenhouse effect—but the jury is not in, the final verdict has yet to be pronounced. And, all the while, we are plagued by the incompleteness of the early rock record. Hard as we may try, we can discover only that fossil evidence which still exists for us to find.

Although we are therefore mightily constrained in our efforts to document, and decipher, the Precambrian history of life, we can take one more step toward understanding—we can make an educated guess. This is not as wild and woolly an exercise as it might first appear, because a really good guess might cause us to ask new questions that lead to increased knowledge.

To play this game, we first need to take into account all of what is currently known—each of our guesses must be fully consistent with *all* of the evidence now at hand.

To these currently known facts, we have to add those new pieces of the puzzle that we have sound reasons to believe are likely to be discovered in the not too distant future. Twenty years ago, this would have been a fruitless exercise. No one then knew what the Precambrian fossil record would reveal. But, now, with two decades of experience behind us, this is not such a daunting task. Future discoveries are certain to provide important new insight, but the overall picture is likely to remain largely unchanged.

And, finally—and this is the really difficult part of the game—we have to add to this mix a good guess as to what the early fossil record *really* would be like if it all still existed for us to study, if all of the evidence that once had been deposited had actually survived to the present day. Why is such a "guesstimate" so difficult? Simply because it takes us out of the realm of testable science; there is no way to check, to know for certain, whether our guesses are really good.

Why is this game worth playing? Because, probably without even realizing it, each of us has a tendency to assume that the evidence available to us, now, represents the whole truth. But the available facts almost always tell only a part of the real story. Thus the game is useful because it forces us to separate in our minds what we actually know, from what we think will be known, from what we only can hypothesize. Moreover, if we are clever at playing this game, some aspect of the resulting educated guess may lead us to ask questions that are testable in the real rock record, questions that will lead to new facts and new understanding.

Where does all this leave us? Well, the results of one such educated guess are summarized in Figure 2.16. The two diagrams on the left of the figure illustrate especially well what can be learned from this sort of approach. Note the differences between the "known distributions" and the "estimated distributions" of both microbial stromatolites and free-living prokaryotes. What these plots basically suggest is that what is known, now, is only a small fragment of the total picture. Of course, it remains to be determined whether or not this particular educated guess is a good one; it will be either confirmed or refuted as new data are amassed. What does seem clear, however, is that much remains to be learned—the whole story of the early history of life has yet to be told!

Take-Home Lessons

Finally, there are two additional observations—useful "take-home lessons"—to be gained from this discussion of the earliest records of life on earth:

FIGURE 2.16

Schematic diagram summarizing the known and estimated (educated guess) abundances of microbial stromatolites, and of fossil prokaryotes, eukaryotic phytoplankton, metaphytes (multicellular algae and higher plants), and metazoans (invertebrate and vertebrate animals), over Precambrian and Phanerozoic time.

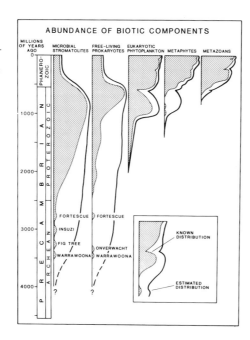

1. Life and the environment are intertwined, intermeshed. One irrevocably affects the other. Nature is not compartmentalized into the familiar academic disciplines. To understand nature, how it is structured, how it has changed over time, knowledge of biology alone is not enough. Nor of geology alone. Nor of atmospheric sciences. Nor of chemistry or biochemistry. Not any one of these disciplines, by itself, is sufficient. Effective investigation of the history of life requires, absolutely demands, a multidisciplinary approach.

2. It will be a true triumph when an understanding of the total history of life is ultimately achieved—when it is finally understood how all of the pieces fit into the vast evolutionary puzzle. But this is a success that has yet to be celebrated. Certainly the longest, and probably the most fundamental phase of this story, that of the Precambrian history of life, is only now coming into focus. The future of the field is bright; the studies will be arduous, but they are certain to prove rewarding; the fruit is ripe for the plucking:

> And pluck, til time
> and times are done . . .
> The golden apples of the sun.
>
> William Butler Yeats

FURTHER READING

History of Precambrian Paleobiology

Cloud, P. 1983. Early biogeologic history: The emergence of a paradigm. In J. W. Schopf (Ed.), *Earth's Earliest Biosphere, Its Origin and Evolution* (Princeton, NJ: Princeton Univ. Press), pp. 14–31.

Schopf, J. W. 1992. Historical development of Proterozoic micropaleontology. In J. W. Schopf and C. Klein (Eds.), *The Proterozoic Biosphere, A Multidisciplinary Study* (New York: Cambridge Univ. Press), Section 5.2 (in press).

The Precambrian Rock Record

Garrels, R. M., and Mackenzie, F. T. 1971. *Evolution of Sedimentary Rocks* (New York: Norton), 397 pp.

Lowe, D. R. 1992. Major events in the historical development of the Precambrian earth. In J. W. Schopf and C. Klein (Eds.), *The Proterozoic Biosphere, A Multidisciplinary Study* (New York: Cambridge Univ. Press), Section 2.7 (in press).

Modern Cyanobacteria

Carr, N. G., and Whitton, B. A. (Eds.). 1982. *The Biology of Cyanobacteria*, Botanical Monographs, Vol. 19 (Oxford: Blackwell), 688 pp.

Fay, P., and Van Baalen, C. (Eds.). 1987. *The Cyanobacteria* (New York: Elsevier), 543 pp.

Packer, L., and Glazer, A. N. (Eds.). 1988. *Cyanobacteria*, Methods in Enzymology, Vol. 167 (New York: Academic Press), 915 pp.

Modern and Fossil Stromatolites

Pierson, B. K., Bauld, J., Castenholz, R. W., D'Amelio, E., Des Marais, D. J., Farmer, J. D., Grotzinger, J. P., Jørgensen, B. B., Nelson, D. C., Palmisano, A. C., Schopf, J. W., Summons, R. E., Walter, M. R., and Ward, D. M. 1992. Modern mat-building microbial communities: A key to the interpretation of Proterozoic stromatolitic communities. In J. W. Schopf and C. Klein (Eds.), *The Proterozoic Biosphere, A Multidisciplinary Study* (New York: Cambridge Univ. Press), Chapter 6 (in press).

Walter, M. R. (Eds.). 1976. *Stromatolites,* Developments in Sedimentology 20 (New York: Elsevier), 790 pp.

Precambrian Organic and Isotopic Biogeochemistry

Hayes, J. M., Des Marais, D. J., Lambert, I. B., Strauss, H., and Summons, R. E. 1992. Proterozoic biogeochemistry. In J. W. Schopf and C. Klein (Eds.), *The Proterozoic Biosphere, A Multidisciplinary Study* (New York: Cambridge Univ. Press), Chapter 3 (in press).

Hayes, J. M., Kaplan, I. R., and Wedeking, K. W. 1983. Precambrian organic geochemistry, preservation of the record. In J. W. Schopf (Ed.), *Earth's Earliest Biosphere, Its Origin and Evolution* (Princeton, NJ: Princeton Univ. Press), pp. 93–134.

Schidlowski, M., Hayes, J. M., and Kaplan, I. R. 1983. Isotopic inferences of ancient biochemistries: Carbon, sulfur, hydrogen, and nitrogen. In J. W. Schopf (Ed.), *Earth's Earliest Biosphere, Its Origin and Evolution* (Princeton, NJ: Princeton Univ. Press), pp. 149–186.

Archean Stromatolites and Microfossils

Schopf, J. W. 1992. Paleobiology of the Archean. In J. W. Schopf and C. Klein (Eds.), *The Proterozoic Biosphere, A Multidisciplinary Study* (New York: Cambridge Univ. Press), Section 1.5 (in press).

Schopf, J. W., and Packer, B. M. 1987. Early Archean (3.3-billion to 3.5-billion-year-old) microfossils from Warrawoona Group, Australia, *Science 237:* 70–73.

Schopf, J. W., and Walter, M. R. 1983. Archean microfossils: New evidence of ancient microbes. In J. W. Schopf (Ed.), *Earth's Earliest Biosphere, Its Origin and Evolution* (Princeton, NJ: Princeton Univ. Press), pp. 214–239.

Walter, M. R. 1983. Archean stromatolites: Evidence of the earth's earliest benthos. In J. W. Schopf (Ed.), *Earth's Earliest Biosphere, Its Origin and Evolution* (Princeton, NJ: Princeton Univ. Press), pp. 187–213.

Evolution of the Atmosphere and Oceans

Holland, H. D. 1978. *The Chemistry of the Atmosphere and Oceans* (New York: Wiley-Interscience), 351 pp.

Holland, H. D. 1984. *The Chemical Evolution of the Atmosphere and Oceans* (Princeton, NJ: Princeton Univ. Press), 582 pp.

Klein, C., Beukes, N. J., Holland, H. D., Kasting, J. F., Kump, L. R., and Lowe, D. R. 1992. Proterozoic atmosphere and ocean. In J. W. Schopf and C. Klein (Eds.), *The Proterozoic Biosphere, A Multidisciplinary Study* (New York: Cambridge Univ. Press), Chapter 4 (in press).

Walker, J. C. G. 1977. *Evolution of the Atmosphere* (New York: Macmillan), 318 pp.

Walker, J. C. G., Klein, C., Schidlowski, M., Schopf, J. W., Stevenson, D. J., and Walter, M. R. 1983. Environmental evolution of the Archean-Early Proterozoic earth. In J. W. Schopf (Ed.), *Earth's Earliest Biosphere, Its Origin and Evolution* (Princeton, NJ: Princeton Univ. Press), pp. 260–290.

Hypobradytelic Evolution of Proterozoic Prokaryotes

Schopf, J. W. 1992. Proterozoic prokaryotes: Affinities, geologic distribution, and evolutionary trends. In J. W. Schopf and C. Klein (Eds.), *The Proterozoic Biosphere, A Multidisciplinary Study* (New York: Cambridge Univ. Press), Section 5.4 (in press).

Schopf, J. W. 1992. Tempo and mode of Proterozoic evolution. In J. W. Schopf and C. Klein (Eds.), *The Proterozoic Biosphere, A Multidisciplinary Study* (New York: Cambridge Univ. Press), Section 13.3 (in press).

Proterozoic Eukaryotic Microfossils

Mendelson, C. V., and Schopf, J. W. 1992. Proterozoic and Early Cambrian acritarchs. In J. W. Schopf and C. Klein (Eds.), *The Proterozoic Biosphere, A Multidisciplinary Study* (New York: Cambridge Univ. Press), Section 5.5 (in press).

Proterozoic Microfossil Diversity

Schopf, J. W. 1991. Collapse of the Late Proterozoic ecosystem. *S. African J. Geol.* (in press).

Schopf, J. W. 1992. Patterns of Proterozoic microfossil diversity: An initial, tentative analysis. In J. W. Schopf and C. Klein (Eds.), *The Proterozoic Biosphere, A Multidisciplinary Study* (New York: Cambridge Univ. Press), Section 11.3 (in press).

CHAPTER 3

EVOLUTION OF THE EARLIEST ANIMALS

■

Bruce Runnegar†

■

INTRODUCTION

Metazoans (multicelled animals) appear abruptly in the fossil record at the close of the Precambrian. Although this striking event has been obvious since early in the study of historical geology and has attracted a good deal of scientific attention during the last decade, several fundamental aspects of the early history of the Metazoa remain unclear. For example, are metazoans a **monophyletic group** derived from a single kind of unicelled organism? If so, are metazoans descended from ciliated unicells, from other kinds of unicells, or perhaps from multicellular eukaryotes such as fungi or plants? Was the Precambrian history of the Metazoa relatively long (for example, 500 million years) or relatively short (less than 100 million years)? Were late Precambrian soft-bodied organisms, collectively known as the Ediacara fauna, fundamentally different from Paleozoic and younger metazoans as Seilacher (1989) has suggested; and, if so, was there a mass extinction at the end of the Precambrian? And finally, how and when did the major higher taxa (phyla) of living animals evolve?

Until recently, it would have been difficult to seek precise answers to many of these questions. Studies of comparative anatomy enabled 19th-century biologists to classify animals effectively into phyla, but the evolutionary relationships of the different phyla remained obscure because there were too few shared morphological characters available for analysis. The invention of the electron microscope allowed a wealth of new ultrastructural data to be applied to the problem, but it was the subsequent introduction of rapid and efficient methods of sequencing protein and nucleic acid (DNA and RNA) molecules which ultimately invigorated the field by providing access to the vast amount of historical information that is stored in the genomes of all living creatures. In essence, this information is digital in character because proteins and nucleic acids are linear polymers of either 20 kinds of amino acids or 4 kinds of nucleotides. The stored information can therefore be used

†Department of Earth and Space Sciences, Molecular Biology Institute, and Institute of Geophysics and Planetary Physics, University of California, Los Angeles, California 90024 USA

quantitatively to assess the degree of relatedness (the recency of common ancestry) among representatives of the living animal phyla. Several alternative methods of numerical analysis are commonly employed (Li and Graur, 1990); the final product is a **dendrogram,** or "tree" which displays the purported genealogical relationships in a graphical fashion (Figure 3.1).

A second important contribution from molecular biology has come from recent work on the molecular genetics of animal development. The discovery, first in *Drosophila,* and then in other animals including vertebrates, of related (homologous) regulatory genes that control the formation of body segments (Lewis, 1978; Wilkinson et al., 1990) has led to an appreciation that the serial repetition of body parts is an ancient feature of animal design (Jacobs, 1990; Saint, 1990). This new knowledge of the biochemistry of development promises to provide a better understanding of the fundamentals of animal construction and the modifications that may result from changes in the rate of growth or the timing of critical events in the life cycle. It has already led, albeit indirectly, to an innovative hypothesis for the origin of arthropods with two-branched (biramous) limbs from those having only one pair of unbranched legs per body segment (Emerson and Schram, 1990).

There have been equally significant recent discoveries in invertebrate paleontology. Stimulated in part by an international endeavor to define the Precambrian-Cambrian boundary for the global geological time scale, studies of Cambrian fossiliferous strata in Siberia, China, Europe, and Australia have yielded an unexpected range of well-preserved phosphatic microfossils. Most are either tiny shells or the disarticulated components (sclerites) of a protective armor comprised of many parts (Bengtson et al., 1990). These microfossils have become familiar as the "small shelly fossils." Most are original phosphatic skeletons or phosphatic copies of other kinds of hard parts, but some are unmineralized animals that have been coated with calcium phosphate soon after death (Müller and Walossek, 1985, 1987, 1988; Walossek and Müller, 1990). The latter belong to a class of fossiliferous deposits known as *Lagerstätten* (Seilacher et al., 1985), each of which is famous for its exceptional preservation—frequently of soft parts—and for the window it provides to an otherwise unseen world.

Studies of two successive clusters of *Lagerstätten* have greatly expanded our knowledge of early animal life. The older set is latest Precambrian **(Vendian)** in age and is collectively known as the **Ediacara fauna** from its discovery, in 1946, at a disused base-metal mine called Ediacara in the northern Flinders Ranges of South Australia (Sprigg, 1988). This distinctive suite of impressions of soft-bodied organisms in sandstones or siltstones has now been found in many parts of the world; the most important sites are in southern Africa (Namibia), Newfoundland, Russia (the White Sea coast and the Ural Mountains), the Ukraine (Podolia), and South Australia (Glaessner, 1984).

The younger set of *Lagerstätten* includes the celebrated Middle Cambrian Burgess Shale of British Columbia (Gould, 1989; Whittington, 1985). Other similar but less spectacular occurrences are known from a number of Middle and Late Cambrian localities in North

FIGURE 3.1

Metazoan tree derived from two trees calculated by Lake (1990) from 18S rRNA sequences using the method known as "evolutionary parsimony." Numbers in brackets refer to the number of species used for each branch (redrawn from Patterson, 1990).

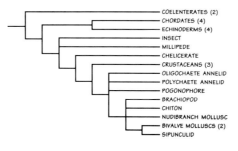

America (Collins et al., 1983; Conway Morris, 1985a; Conway Morris and Robison, 1986), but the real rivals of the Burgess Shale are the newly discovered Chenjiang fauna of south China (Chen et al., 1991) and the equally novel Buen fauna of north Greenland (Conway Morris and Peel, 1990). Both are Early Cambrian in age and neither is, as yet, fully explored.

It is not difficult to trace most of the extant animal phyla back to the Early Cambrian. At that time they coexisted with other animals that seem very strange to us because they have no living descendants. Such creatures are sometimes wrongly placed in living phyla, but they are more often recognized as being difficult to classify and are therefore treated as either *incertae sedis* (Latin, uncertain place) or as members of the *Problematica* (Bengtson, 1986). In this context, it is useful to distinguish between organisms belonging to a **crown group**—those that are descended from the last common ancestor of all living members of a clade—and those that belong to various **stem groups** and their extinct side branches (Figure 3.2). It is difficult to apply traditional classificatory categories (phylum, class, order, etc.) to such short-lived experiments (Craske and Jefferies, 1989). Glaessner (1984, p. 135) put it succinctly: "In historical perspective, failures are not phyla."

Our confidence in the evidence for the existence of early members of the living animal phyla evaporates as we move backward in time into the latest Precambrian (Seilacher, 1989). To some extent, this uncertainty is a reflection of the lower quality and quantity of Vendian fossils compared with their Cambrian counterparts, but it is also due to the fact that the organisms of the Ediacara fauna are odd by any standards. As Seilacher (1989) has pointed out, it is no longer possible to rely on assumed homologies with living animals when attempting to interpret these enigmatic fossils. Instead, the paleobiologist is forced to approach them from "first principles" using only logic and experiment. It is the kind of analysis that would need to be undertaken if we were to find fossils of extinct multicellular life on Mars, and it is sobering to realize how difficult a task it would be.

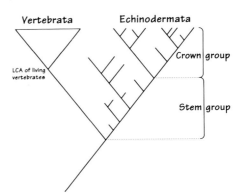

FIGURE 3.2

Hypothetical tree illustrating the difference between a stem group and a crown group. A crown group contains all descendants of the last common ancestor (LCA) of a monophyletic living group (e.g., the Vertebrata). On the echinoderm side, all fossil forms that postdate the split between the echinoderms and the vertebrates, but which lived before the last common ancestor of all living echinoderms, belong to a stem group. Stem group organisms are those that were acquiring the features of the crown group; they do not fit comfortably into modern higher taxa (phyla, classes, orders, etc.); redrawn with modifications from Paul and Smith (1984).

The quest for fossils of the first animals has yielded a variety of structures from sedimentary rocks as old as about 2000 Ma (Kauffman and Steidtmann, 1981). Many of these objects have been shown to be nonbiological in origin, and therefore pseudofossils rather than true fossils, but a few are obviously the remains of once-living organisms, which may or may not have been multicellular animals. For example, *Grypania spiralis* (Figure 3.11) was a coiled, spaghettilike creature up to half a meter in length that is found in ~1300-Ma-old shales and slates in Montana, China, and India. Was it a large unicell—perhaps with many nuclei—or was it multicellular? How did it obtain its food and its energy? How did it reproduce and grow? How is it related to animals, fungi, and plants? These are the kinds of questions we need to ask about all Precambrian fossils that might represent the remains of early animals.

The following discussion deals with some of these matters. We begin with a brief review of the phyla of living animals, their relationship to other multicellular organisms, and the paleontological evidence for their time of appearance in the fossil record. It will then be possible to consider, in a more informed way, the nature of the organisms that constitute the Ediacara fauna. We then examine the evidence for the existence of other early animals, and finally, we attempt to put the evolution of animals into the broader context of the evolution of the biosphere and its effects on planet earth.

THE METAZOAN CLADE

Living animals come in a great variety of shapes and sizes, but they share a few basic properties which indicate that all animals are descended from a single species that lived some time during the later Precambrian. These shared derived characters include a multicellular body formed from different kinds of cells; the ability to manufacture the protein collagen; a reproductive cycle with gametes (sex cells) produced by meiosis; and, in all animals except sponges, the existence of a nervous system composed of neurons (Ax, 1989). These common characters, which are unique to animals, allow all living and fossil animals to be placed in a single monophyletic taxon called the **Metazoa.**

The taxonomic integrity (monophyly) of the Metazoa has been tested recently by analyses of the nucleotide sequences of 18S ribosomal RNA molecules from a number of distantly related living animals. When the data were first obtained they were analyzed using a procedure that measures the evolutionary distance between all pairs of species included in the study. This method suggested that the metazoans are not monophyletic; the cnidarians (*Hydra* and a sea anemone) clustered with maize, yeast, and a ciliate protist rather than with the higher animals (Field et al., 1988). However, other analyses, using methods based on the concepts of **parsimony** (the smallest number of changes required to produce the observed result) and **cladistics** (the recognition of relationships by means of shared derived characters), have confirmed the monophyletic nature of at least the **Eumetazoa** (all animals except sponges; Lake, 1990; Patterson, 1989). Because sponges have collagen genes that are homologous to those of echinoderms and vertebrates (Exposito and Garrone, 1990), and because collagen is a protein which is confined to animals, it is almost certain that all metazoans belong to a monophyletic clade.

Living animal species represent only a small sample of those that have existed in the past. The same principle applies at higher taxonomic levels, so we must expect to find evidence in the fossil record of extinct genera, families, orders, classes, and even phyla. However, the recognition of extinct taxa is complicated by the fact that differences between the branches of the tree of life become smaller as we move further back in time. Moreover, there are generally more branches low down on the tree, so the distinctiveness of any particular branch does not become apparent until the tree is more fully grown. It is therefore convenient to start with living animals when constructing or reviewing a classification. Living animals are exactly contemporaneous, so none can be the ancestor of another, and living animals are normally better known than fossils (see Craske and Jefferies, 1989, and Walossek and Müller, 1990, for a full discussion of this matter).

Sponges (Porifera) are the **sister group** of all other living metazoans (Figure 3.3; Ax, 1989). This means that all other metazoans are descended from a single species that was also the ancestor of the sponges and their extinct relatives, the Archaeocyatha. Because this ancestral species is unlikely to have been either a sponge or an archaeocyath, its time of existence must predate the appearance of sponges and/or archaeocyaths in the fossil record. Similarly, the coelenterates (Cnidaria plus Ctenophora) are the sister group to all other living higher Metazoa (**Bilateria;** Ax, 1989); their origin must predate the first appearance of bilaterally symmetrical animals having three body layers (ectoderm, mesoderm, and endoderm).

Within the Bilateria there are two possibilities (Ax, 1989). Either the flatworms (Platyhelminthes) form a sister group to the **Eubilateria** (Annelida, Arthropoda, Echinodermata, Mollusca, Chordata, etc.), or a one-way gut evolved twice—once in the line leading to annelids, arthropods, and molluscs, and once in the line leading to echinoderms and chordates. In the second alternative the position of the **Tentaculata** (Phoronida, Brachiopoda, and Bryozoa) is ambiguous, but the available molecular data suggest that the tentaculates belong with the annelids, arthropods, and molluscs rather than with the **Deuterostomia** (Chaetognatha, Echinodermata, and Chordata). Most of the remaining living animal phyla are either parasitic or microscopic and, therefore, are of little paleontological significance, or are allied to the annelids, arthropods, and molluscs (Echiurida, Onychophora, Pogonophora, Priapulida, Sipuncula). Nematodes probably lie between flatworms and the higher Eubilateria (Figure 3.3).

The Metazoa in a Wider Context

The relationships of the three major extant groups of multicellular organisms—animals, plants, and fungi—is complicated by the fact that the fungal "kingdom" may not be monophyletic (Smith, 1989). In contrast, as discussed in Chapter 4, there is good morphological and molecular evidence that plants comprise a monophyletic clade descended from charophycean green algae (Devereux et al., 1990; Manhart and Palmer, 1990), and there is a limited amount of molecular evidence suggesting that animals and

FIGURE 3.3

Consensus tree for the principal modern metazoan phyla. Each of these phyla except the Bryozoa has an observed origin (black bars) or inferred origin (shaded bars) prior to the beginning of the Phanerozoic. The order of appearance of the phyla, deduced from comparative anatomy and molecular sequence data, is shown at the bottom of the diagram. The length of the Precambrian (pre-Phanerozoic) history of the Metazoa is unknown.

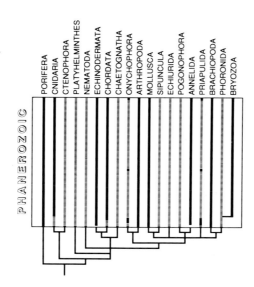

plants are sister groups derived from a common, perhaps unicellular, ancestor (Guoy and Li, 1989; Perasso et al., 1989). Thus, the traditional botanical view, in which fungi are more closely related to plants than to animals, is not strongly supported by molecular data at the present time.

THE PHANEROZOIC FOSSIL RECORD

The known fossil record of animals begins in rocks that are about 600 to 550 Ma old, and the subsequent abrupt appearance of many different kinds of animals with mineral skeletons identifies the beginning of an eon of geological time known as the **Phanerozoic.** As discussed in Chapter 2, the Phanerozoic is the younger of two unequal portions of earth history. The older portion is the Precambrian; it lasted from the formation of the earth until approximately 530 Ma ago.

The exact age of the beginning of the Phanerozoic is unknown. Fossils are found in sedimentary rocks; these rarely contain minerals that both formed when the sediment was deposited and also have the right kinds of isotopes used for age determination. It has therefore been necessary to use isotopic dating techniques that are less certain than those applied to igneous and metamorphic rocks. As a result, the assumed age of the beginning of the Phanerozoic (the Precambrian-Cambrian boundary) has fluctuated with the method of analysis. The latest approach—using an ion probe to measure the ratios of uranium and lead ($^{206}Pb/^{238}U$ and $^{207}Pb/^{206}Pb$) in single crystals of zircon from volcanic ash beds—yields ages that are significantly younger than those obtained by other methods (Figure 3.4). Thus the "textbook" age of the Precambrian-Cambrian boundary has recently been reduced from 570–590 Ma to 530–550 Ma. This revision toward a younger age, if correct, is of some importance because the end of the Cambrian appears to be no younger than about 500–510 Ma. Consequently, the Cambrian radiation of the Metazoa may have been compressed into a much shorter interval of time than previously considered possible—perhaps as little as ten million years.

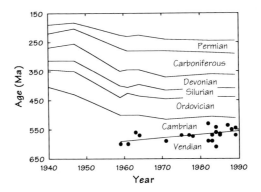

FIGURE 3.4

History of estimates of the true ages of the Paleozoic periods and the Precambrian-Cambrian boundary over the past 50 years. The estimated age of the beginning of the Cambrian has been getting younger (line fitted to 21 separate estimates) as the ages of the other periods of the Paleozoic have increased. Consequently, the "Cambrian explosion" is becoming compressed into substantially less time than was previously thought.

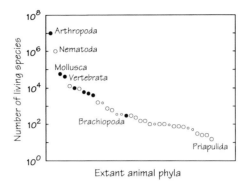

Estimates of the number of living species of the metazoan phyla plotted in order of abundance. Phyla shown as solid circles have members with mineral skeletons; the small circles represent phyla comprised of microscopic organisms. Generally speaking, mineralized, abundant, and/or large animals are the ones that are likely to be fossilized. The Priapulida, which are found in several of the famous fossil *Lagerstätten,* are an exception to the rule.

About a quarter of the living metazoan phyla contain species with mineral skeletons (Figure 3.5); these animals have a good fossil record throughout much of the Phanerozoic. Other living phyla appear sporadically as soft-bodied fossils in the exceptional deposits called *Lagerstätten,* although not necessarily in proportion to their present abundance (Figure 3.5). The *Lagerstätten* are particularly useful in the early Paleozoic, for they allow us to trace some wholly soft-bodied phyla, such as the Priapulida, back to the Early Cambrian (Conway Morris, 1977; Conway Morris and Robison, 1986; Hou and Sun, 1988). In other cases, the observed stratigraphic range of a sister group may be used as a proxy for the longevity of a phylum that has a poor or nonexistent fossil record. For example, anatomical, embryological, physiological, paleontological, and molecular evidence identifies the Sipuncula as the extant sister group of the Mollusca (Lake, 1990; Rice, 1985; Runnegar et al., 1975). Consequently, the origin of the Sipuncula must predate the first appearance of the Mollusca in earliest Cambrian strata (Runnegar and Pojeta, 1985).

If this procedure is applied to other pairs of phyla that are known to be closely related to each other (for example, Cnidaria-Ctenophora, Brachiopoda-Phoronida, Arthropoda-Onychophora, Annelida-Pogonophora, and Echinodermata-Chordata), it is possible to infer that most of the living animal phyla have their roots in the Precambrian or earliest Cambrian (Figure 3.3). The same may be true for extinct taxa of phylum or near-phylum grade (Figure 3.6). The best known of these are the Archaeocyatha (?Porifera; Hill, 1972); bizarre echinoderms called "carpoids" (Jefferies, 1990); the Coeloscleritophora (*Chancelloria, Wiwaxia, Halkieria,* and their allies; Bengtson and Conway Morris, 1984; Conway Morris, 1985b; Conway Morris and Peel, 1990); the Conodonta (?Chordata; Aldridge and Briggs, 1986; Dzik, 1986a); the Conulariida (Babcock and Feldmann, 1986); the Hyolitha (Runnegar et al., 1975); the Machaeridia (Dzik, 1986b); the Palaeoscolecida (?Annelida; Conway Morris and Robison, 1986); and a number of unusual animals from the Burgess Shale and other Cambrian *Lagerstätten* (for example, *Anomalocaris, Hallucigenia, Microdictyon, Opabinia,* etc.; Gould, 1989; Whittington, 1985).

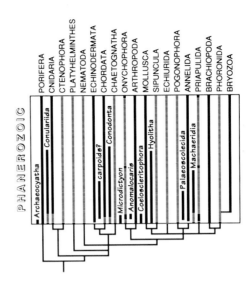

FIGURE 3.6

Consensus tree for the principal modern metazoan phyla, with important extinct groups of phylum or near-phylum grade plotted next to their nearest living relatives. The extinct groups tend to fill the morphological gaps between the modern phyla. Black bars represent groups with a good fossil record; shaded bars are inferred stratigraphic ranges.

THE ENIGMA OF EDIACARA

When Reg. C. Sprigg discovered the saucer-sized type specimen of *Ediacaria flindersi* at Ediacara in March 1946, he immediately identified it as a jellyfish belonging to the cnidarian class Scyphozoa (Sprigg, 1947, 1988). This interpretation was reinforced by the collection of many more discoidal fossils in 1947, and it soon became widely accepted that jellyfish impressions are the most common kinds of fossils to be found at Ediacara (Sprigg, 1949; Wade, 1972a).

In 1957, a schoolboy, Roger Mason, found complex frondlike fossils attached to the centers of disk-shaped structures in late Precambrian rocks of the Leicestershire district of England. They were first considered to be fossil algae, but that idea was set aside when similar fossils were reported from Ediacara and interpreted as the remains of organisms related to living pennatulate cnidarians—sometimes called "sea-pens" (Glaessner, 1959). Very rapidly thereafter, the repertoire of animals recognized at Ediacara was increased by new discoveries. Then in 1961, Martin F. Glaessner published an influential article in *Scientific American* in which he confirmed the Precambrian age of the Ediacaran fossils, demonstrated their global character, and related most to modern animal groups (Glaessner, 1961). This interpretation of the Ediacara fauna has been followed, modified, and extended by other paleontologists who have worked in South Australia (e.g., Gehling, 1988; Jenkins, 1985; Runnegar, 1982a; Sun, 1986; Wade, 1972).

Fossils of late Precambrian soft-bodied organisms were discovered by P. Range and H. Schneiderhöhn in South-West Africa (Namibia) in 1908 and 1914, but they were not reported in the scientific literature until the 1930s. Most of the fossils are the distorted remains of three species, all apparently sessile and frondlike or bag-shaped—*Rangea schneiderhoehni* (Figure 3.7), *Pteridinium simplex* (Figure 3.7), and *Ernietta plateauensis*. They were first thought to be sponges or coelenterates, but in 1970 H. D. Pflug recognized that these Namibian fossils could not easily be placed in any living animal group. He therefore proposed a new extinct phylum, the Petalonamae (and also proposed many unnecessary new generic and specific names) to accommodate them (Pflug, 1970a,b, 1972a,b). This idea was not well received because it was accompanied by complex, even fanciful, reconstructions of the animals which Pflug believed represented a monophyletic radiation of early colonial Metazoa.

In contrast to Glaessner, who referred most of the Ediacaran fossils to modern animal groups, and Pflug, who interpreted some as members of an extinct phylum, a third strategy was adopted by M. A. Fedonkin in the Soviet Union (Fedonkin, 1984, 1985a,b, 1987). He began with the idea that it "would be more promising to consider the Vendian [Ediacara] fauna, not as ancestors of the Phanerozoic fauna, but as descendants of the older Precambrian metazoans, still unknown to us" (Fedonkin, 1985a, p. 12). He therefore used differences in symmetry to identify and categorize major taxonomic groups of Vendian age that, in general, were regarded as extinct higher taxa of phylum- or class-grade. A model for the early history of the Metazoa based on symmetry-related changes in gross body form resulted from this work.

It was A. Seilacher, who, in 1983, invigorated the study of the Ediacaran fossils with the proposal that few, if any, of the organisms can be referred to living phyla or even to the animal kingdom (Seilacher, 1989). Reassessed in this way, it is clear that few of the described taxa can unequivocally be referred to known living or extinct animal groups. For example, some of the discoidal "medusoids" may be the remains of cnidarian jellyfish, but none has the diagnostic cnidarian characters—such as fourfold symmetry—that are visible in fossil jellyfish from younger *Lagerstätten* (Barthel, 1978; Nitecki, 1979). In fact, the

---

FIGURE 3.7

Core members of the Ediacara fauna and Seilacher's "Vendozoa." *Left: Pteridinium simplex* from the coast of the White Sea, Russia, about natural size. The image of the upper part of the specimen has been repeated above (area denoted by arrow) to illustrate that the younger parts of the specimen were exact copies of the older parts (photograph courtesy of M. A. Fedonkin). *Middle:* part of the unnamed "spindle-shaped organism" from Newfoundland, somewhat smaller than natural size (drawing courtesy of A. Seilacher). *Right:* sketch of *Rangea schneiderhoehni* reconstructed from specimens from Namibia (from Jenkins, 1985, republished with permission).

  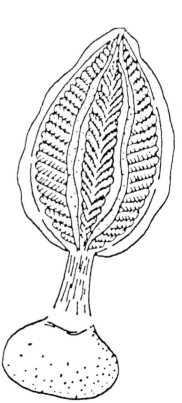

only clear indication of radial symmetry in the discoidal fossils of the Ediacara fauna is threefold (Figure 3.8). Consequently, Fedonkin (1985a) placed these and some other forms with triradial symmetry in a new extinct class of the Cnidaria, which he called the **Trilobozoa.** He pointed out that some populations of modern jellyfish contain individuals which are aberrant in terms of their symmetry, and that it may therefore only be an accident of history that all living jellyfish are based on a tetraradial design. If we were able to "rerun" the Phanerozoic, all living jellyfish might have triradial symmetry and all land vertebrates (tetrapods) might have six legs. For example, there is paleontological evidence suggesting that pentadactyly (the possession of five fingers or toes) is not a trait inherited from the earliest tetrapods but rather is a condition which developed as a result of functional and developmental constraints in populations of animals that had a variable number of digits (Coates and Clack, 1990). There is no a priori reason for believing that five digits is the inevitable outcome of such a natural experiment.

Despite the undoubted value of Seilacher's new approach to this problem, it is possible that some of the assignments of Ediacaran fossils to modern phyla which were made in the past may be correct but unprovable. The recognition of many critical features is hampered by the nature of the preservation of the fossils. It is therefore necessary to try to understand how the soft-bodied organisms were fossilized, and why these kinds of fossils are not found in younger marine sedimentary rocks.

Preservation of the Ediacara Fauna

At Ediacara and other localities in the Flinders Ranges of South Australia, the fossils occur as shallow concave depressions, or as comparable convex elevations, on the bases of

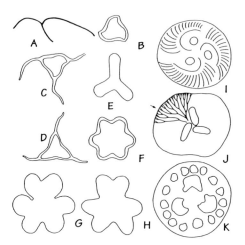

FIGURE 3.8

Sketches of Vendian and Cambrian fossils that have a triradial structure or threefold symmetry. (A) End-on view of the three "wings" of *Pteridinium* as seen on the broken surface of a piece of sandstone, somewhat smaller than natural size; Vendian, Namibia. (B–H) Cross-sections of various species of the microscopic Early Cambrian "small shelly fossil" *Anabarites* from Siberia and Australia. (I–K) Three trilobozoans, *Tribrachidium* (I), *Albumares* (J), and *Skinnera* (K), from the Vendian of Russia and Australia, each about natural size; only some of the branching radial "canals" are shown on *Albumares,* with the arrow pointing to one of three pathways that are free of "canals."

quartz-dominated sandstone beds. The original grains of quartz are well rounded, but the spaces between the grains have been filled with secondary quartz deposited chemically in the form of a mineral cement. Frequently, the beds of sandstone are only a few centimeters thick, and they often have ripple-marked upper and lower surfaces (the latter being casts of underlying ripple marks). Successive beds were separated by a film or layer of finer sediment and, perhaps, by a thin cyanobacterial mat, which J. G. Gehling has suggested could have coated all exposed sediment surfaces in well-lit, nearshore late Precambrian environments. Although the environment has often been reported to have been intertidal (e.g., Glaessner, 1961), recent sedimentological studies favor deposition at a water depth that was below fair weather wave base. Thus the sandstones that cast the soft-bodied organisms are thought to be storm beds, each being the result of a single short-lived, energetic event in an otherwise calm, deep-water environment. The base of each bed therefore provides a "snapshot" of the seafloor at the time it was deposited.

The rock unit that contains the Ediacara fauna is called the Rawnsley Quartzite. It is about 400 m thick in Bunyeroo Gorge in the central Flinders Ranges, and the fossils are found in a 50-m-thick sequence (Ediacara Member) which is finer grained and thinner bedded than the rest of the formation. A few hours experience at Ediacara is sufficient to enable a paleontologist who is new to the area to locate the fossiliferous horizon in Bunyeroo Gorge before finding the fossils. It is therefore clear that the preservation of the soft-bodied organisms at Ediacara, and in other parts of the Flinders Ranges, depends on special local conditions. The upper and lower limits of the occurrences of fossils have no meaning in terms of evolution or extinction; the Ediacara Member is a **taphonomic (preservational) window** which came and went as a result of environmental changes that occurred only in this part of South Australia during the late Precambrian.

One remarkable aspect of the preservation of some kinds of organisms at Ediacara and elsewhere in South Australia is that they occur as concave external molds on the bases of the beds. If the fossils are collected from rock outcrops rather than from loose blocks, it is sometimes possible to recover the two bed surfaces that originally sandwiched the body. In such cases, the lower bed invariably displays a convex cast of the overlying concave external mold. This indicates that sand flowed upward from the underlying bed to fill the cavity occupied by the body. It seems inconceivable that this would happen unless the roof of the cavity was already strong enough to withstand the force of gravity. Whether this strength derives from early cementation of the sand grains (Wade, 1968), or from a property of the organism, is unknown.

In Newfoundland, the Ediacara fauna is found on large bed surfaces exposed by coastal erosion (Anderson and Conway Morris, 1982; Landing et al., 1988; Misra, 1969). At this locality, there are nine abundantly fossiliferous surfaces within a thickness of 50 m of the Mistaken Point Formation, each at the base of a thin volcanic ash layer. As at Ediacara, the preservation of the soft-bodied organisms depends on local conditions—in this case the deposition of volcanic ash. All available evidence suggests that the organisms were buried where they lived and that the environment was a deep-water one—probably below storm wave base. The dominant components of the assemblage are the famous—but as yet unnamed—"spindle-shaped organisms" (Figure 3.7; Misra, 1969) and frondlike fossils that terminate in disk-shaped holdfasts. The long axes of the fronds are aligned in the direction of the current that carried the volcanic ash, but the spindle-shaped organisms are scattered in random orientations on the fossiliferous surfaces. From these observations, Seilacher has concluded that the spindle-shaped organisms grew like prostrate plants or encrusting animals on the seafloor, whereas the fronds *(Charniodiscus)* stood upright in the water. If the water depth was really below storm wave base (≤ 100 m), then the light intensity must have been quite low (<1% of surface value), particularly if Newfoundland was in relatively high latitudes at the time (Kirschvink, 1992).

A third style of preservation occurs in Namibia. Most of the Namibian fossils come from almost flat-lying beds in a region of low topographic relief. They occur at a transition from sandstone (quartzite) to claystone (shale), but are not normally confined to a single bed surface. Many are embedded in sandstone as twisted or folded, randomly oriented casts and molds; the organisms were obviously engulfed by a moving mass of sand prior to their burial away from the place where they lived.

It is hard to find a common environmental factor that would account for the unusual preservation of the Ediacara fauna at each of these three sites. We are therefore forced to the conclusion that either the organisms had special properties which enhanced their chance of being fossilized, or that the postmortem cleanup procedures carried out by other organisms were not as efficient in the late Precambrian as they were in more recent times. However, as each organism found at all of the sites seems to have been buried alive by an influx of sediment without much prior warning, cleanup procedures must have occurred within the sediment rather than on the seafloor. This constraint eliminates most macroscopic carrion feeders and scavengers from consideration. Thus, for some reason which is not yet well understood, it seems that microbial decomposition and decay failed to destroy the buried bodies (or their tough exteriors) before fossilization could occur.

Nature of the Ediacara Fauna

It has become fashionable to regard **taxonomy** (naming and classifying organisms) as an obsolete field of human endeavor. Nothing could be further from the truth; taxonomy is the basic tool of biology on which all other understanding is built. Thus the lack of a secure taxonomy of the Ediacara fauna is a clear symptom of the poor condition of our knowledge of these extraordinary fossils.

For example, the spectacular type specimen of *Mawsonites spriggi* has appeared on postcards, the cover of *Science,* and on the jacket of the second edition of Richard Cowen's book, *History of Life.* What is it? Glaessner and Wade (1966), who named the object for both Sir Douglas Mawson and Reg. C. Sprigg, regarded it as a jellyfish of uncertain affinities. Sun (1986) went further in placing *Mawsonites* in the cnidarian class Scyphozoa. But Seilacher (1989) believed *Mawsonites* to be the complex burrow of a bilateral metazoan. Until this and similar fundamental differences in interpretation (Table 3.1) can be resolved, it will be difficult to reach any consensus about the nature and significance of the Ediacara fauna.

With these difficulties in mind, let us begin from first principles. The one property that distinguishes animals from plants and fungi is mobility. Ironically, therefore, the best evidence for the existence of animals in the Vendian comes in the obscure form of **trace fossils** (tracks, trails, and burrows left by mobile animals in soft sediments or hard substrates).

The presence of bilaterally symmetrical animals is best demonstrated by the trace fossils. Looping or spiraling surface trails up to several millimeters in width from the Ediacara Member (Figure 3.9) and strings of fecal pellets from the Ediacara fauna of the White Sea coast, point to the existence of soft-bodied Bilateria with a well-developed nervous system, anterior-posterior asymmetry, and a one-way gut. A second kind of trace fossil—which has been found in Ediacara, at the White Sea, and in Podolia—consists of closely spaced meanders that seem to have been formed by a small animal that periodically reversed its direction of motion. These kinds of markings are known as "grazing" traces because their pattern corresponds to the most efficient way of processing a two-dimensional area (e.g., plowing a farm paddock) by moving in a one-dimensional manner (e.g., on a tractor). Raup and Seilacher (1969) used a simple computer program to demonstrate that this kind of behavior requires very few "instructions." It is best interpreted as a form of foraging; thus the animals were presumably eating microbial mats that may have covered the seafloor in shallow, well-lit environments.

TABLE 3.1.

Three Alternative Classifications of Some of the Core Members of the Ediacara Fauna (Crosses in Brackets Identify Extinct Higher Taxa).

Australians	M. A. Fedonkin	A. Seilacher
Coelenterata	Coelenterata	Coelenterata
Hydrozoa	Cyclozoa(†)	Psammocorallia(†)
Ovatoscutum	*Ovatoscutum*	*Beltanelliformis*
Scyphozoa	*Beltanelliformis*	
Ediacaria	*Ediacaria*	Trace Fossil
Mawsonites	*Mawsonites*	*Mawsonites*
Anthozoa	Trilobozoa(†)	
Charnia	*Albumares*	
Charniodiscus	*Tribrachidium*	
Rangea		
Petalonamae(†)	Petalonamae(†)	Vendozoa(†)
	Charnia	*Charnia*
	Charniodiscus	*Charniodiscus*
Pteridinium	*Pteridinium*	*Pteridinium*
Phyllozoon	*Phyllozoon*	*Phyllozoon*
	Rangea	*Rangea*
		Ovatoscutum
Annelida	Proarticulata(†)	*Dickinsonia*
Dickinsonia	*Dickinsonia*	*Spriggina*
Spriggina	*Praecambridium*	*Albumares*
Arthropoda	Arthropoda	*Tribrachidium*
Praecambridium	*Spriggina*	
Incertae Sedis		
Albumares		
Tribrachidium		

FIGURE 3.9

Spiral surface trail a few millimeters in width from the Ediacara Member of the Rawnsley Quartzite, Flinders Ranges, South Australia (photograph courtesy of J. G. Gehling; copyright © 1991 by J. G. Gehling, all rights reserved).

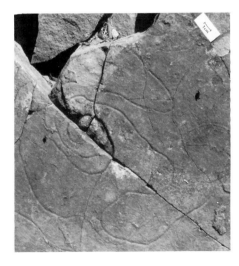

A third kind of structure found in the Ediacara Member (Figure 3.10) was interpreted by Glaessner (1969, 1984) to be the horizontal burrow of a cylindrical animal up to 25 mm in diameter; the fact that some of these structures truncate body fossils (Figure 3.10) was taken as firm evidence for their formation within the sediment after the body had been buried. However, an examination of one of these structures by means of parallel cross-sections shows it to have been yet another kind of body fossil—perhaps the organic tube of a worm or even the elongate stem of a "frond." Thus there is no good evidence for the existence of sizable burrowing animals in deposits that also contain the Ediacara fauna.

This point is important because one of the landmarks of metazoan evolution was the appearance of a fluid-filled body cavity, known as the **coelom.** It has long been debated whether the coelom evolved only once in the history of the Metazoa, or whether it appeared convergently in different lines. As a result of the work of R. B. Clark (1964), it is widely believed that the principal function of the coelom is to enable sizable soft-bodied organisms to burrow in sand or mud. Thus the absence of sizable burrows from rocks of Vendian age has been used as an argument against the existence of coelomate animals (annelids, echiurids, priapulids, sipunculans, etc.) at that time (Bergström, 1989, 1990; Valentine, 1989). However, if the first function of the coelom was merely to increase body size (Runnegar, 1982a), this argument would not apply. Because many kinds of animals (cnidarians, nematodes, annelids, and arthropods) commenced burrowing almost synchronously at the beginning of the Cambrian, it is more likely that the onset of deep burrowing was a response to some general environmental phenomenon such as the appearance of predators (Vermeij, 1987).

The simplest of all Ediacaran body fossils has the form of a round, convex bulge on the base of a bed. Such structures are the ones that are most widespread in space and time, and it is certain that they represent a range of different things preserved in a variety of ways. However, at some sites whole bedding planes are covered with hemispherical blobs (Figure 3.11), each probably the remains of an individual of a single species. They are particularly common in the Vendian of Podolia, Yukon (Narbonne and Hoffman, 1987), and Namibia (Hahn and Pflug, 1988), and are known as *Nemiana simplex* (or sometimes as *Beltanelliformis*). Quite similar structures are found in Phanerozoic rocks (Arai and McGugan, 1968; Orlowski and Radwanski, 1986), and all should probably be identified as the trace fossil *Berguaria hemispherica. Berguaria* appears to be the burrow of a sea anemone (anthozoan cnidarian).

According to Seilacher (1989), *Nemiana* was a member of an extinct group of sessile cnidarians (Psammocorallia) that were similar to sea anemones but, in addition, weighted the middle layer of their bodies (the mesogloea) with sand that was taken in through the

FIGURE 3.10

Sketch of two individuals of *Phyllozoon* (hachured structures) and a tubular body fossil on the base of a sandstone bed, Ediacara fauna, South Australia (courtesy of J. Gehling). The two individuals of *Phyllozoon* are probably superimposed specimens, but they may also have been connected edge to edge during life (somewhat smaller than natural size; tube is 1.5 cm wide).

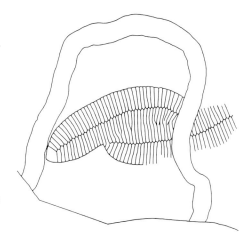

FIGURE 3.11

Precambrian fossils that are difficult to classify. *Left:* a spaghettilike fossil, *Grypania spiralis,* in a rock from India that may be as old as 1300 Ma; somewhat larger than natural size. *Right:* crushed, closely packed dicoidal objects, *Beltanelliformis sorichevae;* Vendian claystone from a borehole near the White Sea, somewhat smaller than natural size (photograph courtesy of M. A. Fedonkin).

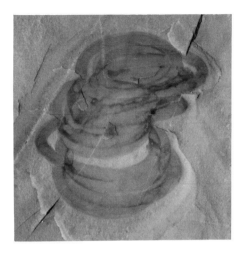 

mouth (Figure 3.12). This seems unlikely because the grain size of the sandstone filling the fossils is the same as the grain size of the bed that buried the organisms. Furthermore, similar fossils *(Beltanelliformis sorichevae)* occur in claystones in which there is no sign of a sand filling (Figure 3.11). If we shelve the idea that *Nemiana* had a sandy skeleton, it is possible to regard the structure as either the burrow or the postmortem filling of an ordinary sea anemone that lived gregariously. However, other interpretations are also possible (Narbonne and Hoffman, 1987), so it is not sensible to place too much weight on such a featureless fossil.

The core members of the Ediacara fauna are those that appear to have a "quilted," or "air mattress" construction (Seilacher, 1989). The most informative of these are the English frondlike form, *Charnia,* the Namibian genus *Pteridinium,* and two genera described originally from Australia—*Dickinsonia* and *Phyllozoon.* Most of these taxa are now known from different parts of the world and they therefore illustrate the global character of the Ediacaran biota. Seilacher (1989) has referred to these organisms as the **Vendozoa** (Vendian animals), but it is also possible that they were not animals at all (Bergström, 1989, 1990).

Seilacher's radical interpretation of these quilted organisms requires them to be members of an extinct monophyletic clade. He therefore shoehorned nearly all other Ediacaran body fossils into the same clade by regarding all observed linear structures (whether concentric, radial, branched, or unbranched) on all fossils as the junctions of fluid-filled compartments isolated by flexible walls. For if this was not the case, the whole thrust of Seilacher's argument would disappear, and we would be left with a set of enigmatic, remotely related early animals.

FIGURE 3.12

Seilacher's concept of Phanerozoic members of the sand corals, or "Psammacorallia" (from Seilacher, 1990, republished with permission). New evidence shows that *Brooksella* is a radial burrow system (A. Seilacher, 1991 personal communication).

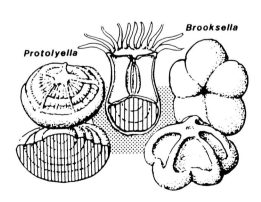

If it could be shown that any two of the quilted taxa were not closely related, but instead had evolved their quilted construction independently, it would be possible to infer that the quilting might have been a response to special environmental conditions that existed at the time. Similarly, if two or three higher taxa could be extracted from the Vendozoa, it becomes less likely that the remainder represents an independent line of "alien life" that is remote from the Metazoa. And finally, if any one of the vendozoans could be shown to have the properties of an animal, it becomes unlikely that the whole group is a separate, nonmetazoan, evolutionary experiment as has been suggested by Seilacher and his supporters. Let us look at each of these tests in turn.

Dickinsonia and *Phyllozoon* are two genera that are at least superficially similar to each other (Figures 3.10 and 3.13). Each is sheetlike, each has a midline and left and right sides, and each is made of compartments, or "segments" that increase in length and number during growth. Both are found in the Ediacara Member so they are exactly contemporaneous. Several hundred specimens of five species of *Dickinsonia* are known (Wade, 1972b). *Dickinsonia* has been found at many places in South Australia, at the White Sea, in Podolia, and in the Ural Mountains; *Phyllozoon*, in contrast, is known only from South Australia (Jenkins and Gehling, 1978).

Phyllozoon (Greek., leaf animal) is the best example of the so-called air mattress construction. The specimens are spread out on bed bases and may overlie one another or be overlain by other kinds of fossils. The compartments of *Phyllozoon* are slightly convex, and the intervening partitions form grooves on the bed base. This is exactly what we would expect if an air mattress had been pressed into soft mud and then was deflated and removed before the marks it left in the mud were cast by plaster or cement poured on top. The compartments on the left and right sides of the organism interlock along a zigzag midline. Growth appears to have occurred by the addition of new compartments, alternately, at the left and right sides of the end (or ends?) of the body. Once a compartment had expanded to the full width of the body, it ceased growing and remained the same size for the rest of the life of the organism.

One particularly fine specimen of *Phyllozoon* collected by J. G. Gehling shows two "individuals" that appear to be joined to one another (Figure 3.10). Each has the usual zigzag midline and evenly spaced compartments, but it looks as if one "individual" grew out of the edge of the other. However, this may be a false impression suggested by a precise, but fortuitous, overlap of the two specimens. If this property of *Phyllozoon* could be confirmed

---

FIGURE 3.13

Species *Dickinsonia* from the Ediacara fauna of South Australia. (A) Fossils of expanded and contracted individuals of *Dickinsonia costata*, each in about the same stage of growth; the larger specimen is about 13 cm in length; note the contraction zone around the large individual and also the small specimen of *Parvancorina* (at arrow). (B) Rubber cast of a lower surface of small individual of *Dickinsonia costata*, about natural size; note the lack of a ridge at the midline. (C) Rubber cast of a folded specimen of *Dickinsonia tenuis*, somewhat smaller than natural size.

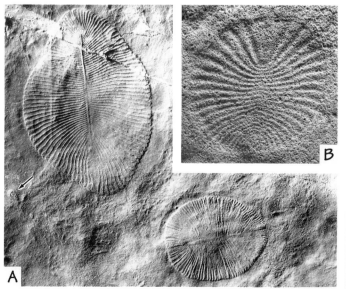

by the discovery of other similar specimens, it would, or course, make *Phyllozoon* even more unlike the other Vendozoa.

The best-known species of *Dickinsonia, Dickinsonia costata,* grew to about the size of a small plate, remaining roughly circular in plan view throughout its growth. As it grew, it increased the number of "segments" from fewer than 5 to more than 70. In contrast to *Phyllozoon,* the upper and lower surfaces of *Dickinsonia* are not identical; on the lower (ventral) surface, the ridges that mark the edges of the "segments" are continuous from one edge to the other (Figure 3.13B), whereas on the upper (dorsal) surface the ridges are interrupted at the midline (Figure 3.13A). There is no evidence that *Dickinsonia* was asymmetrical from left to right as some have claimed; the continuity of the ridges on the ventral surface (Figure 3.13B), and the equal number of ridges on the left and right sides of the dorsal surface (Figure 3.13A, large specimen), show that suggestions left/right of asymmetry are erroneously based on imperfect or distorted specimens. Equally, Seilacher's (1989) assertion that there is no difference between the two ends of *Dickinsonia* is seen to be untrue if we compare the sizes of the first and last intersegmental ridges on any well-preserved specimen (Figure 3.13B). Thus it is clear that *Dickinsonia costata* is bilaterally symmetrical across the body and asymmetrical down the midline. *Phyllozoon* is asymmetrical in both directions.

In South Australia, *Dickinsonia costata* most commonly occurs as concave molds on the bases of beds. If rubber casts are made from these molds, we see a copy of the object that was molded by sand as it lay on the seafloor (Figure 3.13B). Invariably, the junctions between the "segments" are preserved as ridges, and the "segments" themselves, as depressions. The standard explanation for this structure is that the original topography has been inverted; the fluid-filled compartments have collapsed to leave depressions, and the intervening ridges are the tough or stiff materials that formed the walls between the compartments. If this were true, *Dickinsonia costata* would, indeed, have an air mattress construction.

An extraordinary specimen of *Dickinsonia* from Ediacara suggests that this interpretation may be wrong. In this specimen, the bed base has recorded two superimposed views of the organism. The first-formed view (as indicated by overlapping relationships) is a cast of an impression of the lower surface of the body; the later view is a mold of the upper surface. Both upper and lower images have projecting ridges and depressed "segments." Thus the inversion of topography differentiating *Phyllozoon* from *Dickinsonia* may be original, not secondary.

A single individual of *Dickinsonia* was imaged twice in this extraordinary specimen from Ediacara because the creature contracted just prior to being buried (it cannot have moved after being buried or the motion would have smeared the two separate images). This is another property of *Dickinsonia* that distinguishes it from *Phyllozoon.* It turns out that many specimens of *Dickinsonia* are surrounded by smooth areas from which the organism had contracted prior to, or just after, burial (Figure 3.13A, large specimen). Obviously, expansion marks would not be recorded in this way because the organism would override its former position. Evidence of the expansion of *Dickinsonia* is therefore not observed directly, but expansion is inferred to have been just as possible (Runnegar, 1982a; Wade, 1972b). A famous slab of Rawnsley Quartzite from Brachina Gorge in the central Flinders Ranges has two different-sized specimens of *Dickinsonia costata* on its lower surface (Figure 3.13A). Mary Wade (1972b) made the important observation that, because both specimens have approximately the same number of "segments," they must represent expanded and contracted states of individuals of a comparable age.

Another well-known feature of *Dickinsonia* is its method of growth. New "segments" were added at one end, normally taken to be posterior. Newly formed "segments" were tiny at first, and each grew continuously as the whole organism increased in radius. Thus each "segment" gradually moved from a posterior to an anterior position during the growth of the organism. Unlike *Phyllozoon* and *Pteridinium* (Figure 3.7), the whole organism grew continuously, so none of the early ontogeny is visible in the fully grown body. Thus, although *Phyllozoon* is superficially similar to *Dickinsonia costata,* it differs from *Dickinsonia* in fairly fundamental ways, namely in its symmetry, construction, ontogeny and, presumably, in its mobility. These differences draw into question the whole concept of the Vendozoa as an extinct group of closely related organisms.

On the other hand, *Pteridinium* and *Charnia* seem to be closer to *Phyllozoon* than they are to *Dickinsonia,* and they might legitimately be placed in the same higher taxon as *Phyllozoon.* However, the threefold symmetry of *Pteridinium* may also indicate a relationship with *Albumares, Tribrachidium,* and the Early Cambrian "small shelly fossil" *Anabarites* (Figure 3.8). Thus it is possible to find evidence that appears to support the taxonomic integrity of the Vendozoa, as well as evidence that suggests the opposite. The simplicity of the fossils, plus the fact that they are only impressions of soft parts, makes their interpretation very difficult.

For this reason, the most complicated species, *Spriggina floundersi* (Figure 3.14), may be the most informative. *Spriggina* has the same kind of symmetry as *Dickinsonia* (left/right bilateral symmetry, anterior/posterior asymmetry), and it grew in the same way as *Dickinsonia.* Traditionally, it has been interpreted as either an polychaete annelid or a primitive trilobite-like arthropod (e.g., Glaessner, 1984). Once again, Seilacher (1989) has confounded the traditionalists by turning *Spriggina* on its end so that the worm's "head" becomes the frond's "holdfast" (Figure 3.14).

For the moment, *Spriggina* is the star of the debate. Its affinities are critical to an understanding of the time of origin and early history of segment-controlling (homeobox) genes (Jacobs, 1990; Saint, 1990) and to discussions of the early evolution of living (Valentine, 1989) and extinct (Dzik and Krumbiegel, 1989) segmented Metazoa. It is frustrating that no single piece of evidence is available to confirm or deny *Spriggina's* proposed place in the animal world; at the moment, all the evidence we have is either equivocal or circumstantial.

The Garden of Ediacara?

Mark A. S. McMenamin (1986) coined this cliché to promote the idea that the quilted animals of the Ediacara fauna may have housed unicellular photosynthetic algae like those

---

FIGURE 3.14

Two views of a natural sandstone mold of *Spriggina floundersi,* Ediacara fauna, South Australia. In Seilacher's radical interpretation (left), the "head" is the holdfast of a "frond"; in the traditional view (right), the animal is a segmented "worm" with a well-developed "head." Note also that the reversal of the image converts the view of a concave mold (left) to a view of a convex cast (right); fossil is 4 cm in length.

found in many of the sessile animals of modern tropical reefs (e.g., corals, sponges, clams, and foraminiferans). This idea has a good deal of merit because most of the core members of the Ediacara fauna show no trace of a mouth, gut, or anus, and may well have had to obtain their energy from substances taken in through the body wall. If this were the case, the energy source could have been particulate food (e.g., plankton), dissolved organic matter (e.g., amino acids), or even photons (particles of light). If food or dissolved organic matter was the energy source, the organisms would be regarded as **heterotrophs,** but if the organisms used only solar energy in the form of photons, they would be described as **autotrophs.** It is interesting that the boundary between an autotroph and a heterotroph does not coincide exactly with the line between plants and animals; for example, some soft corals have never been observed to feed, and it is assumed that they depend entirely on symbiotic algae (zooxanthellae) for their nutrition (Goreau et al., 1971). They are, therefore, both animals and autotrophs.

Unfortunately, the "Garden of Ediacara" is a hypothesis that is difficult to test. It may be possible to use measurements of the stable isotopes of carbon, oxygen, and nitrogen to see whether Paleozoic reef corals housed algal endosymbionts, but this kind of analysis depends on a source of material that was produced by the animal during its life. As the Ediacaran fossils are nothing more than impressions of soft bodies in sandstones and shales, there is none of the original body available for analysis. It is therefore necessary to resort to indirect evidence to test this useful hypothesis.

Sunlight is reduced to about 0.1% of its peak intensity at the surface in seawater at a depth of about 130 m. Below this depth, photosynthesis is generally thought to be impractical, although a few species of coralline red algae occur below 250 m in the tropics (Littler et al., 1985). So, if the "fronds" and "spindle-shaped organisms" which are found in Newfoundland lived on the seafloor below storm wave base ($\sim$100 m), it is unlikely (but not impossible) that they housed photosynthetic algal symbionts.

Another suggestion made by Seilacher (1989) is that the animals of the Ediacara fauna might have obtained their food by a process known as **chemosymbiosis.** Not much attention had been paid to animals that operate in this way until the first deep-sea thermal "oases" were found in February 1977, by scientists on board the research vessel *Knorr* anchored above the axis of the Galápagos rift valley, east of the Galápagos Islands (Corliss and Ballard, 1977). The animals discovered at that time (e.g., the giant tube-dwelling pogonophore, *Riftia pachyptila*) have now been found to be associated with a large number of modern (and some fossil) hydrothermal vents, including some in the intertidal zone of California. The principal product of the deep-sea vents is hot water containing reduced gases (methane, CH_4, and hydrogen sulfide, H_2S); microorganisms living in the vent waters—or *within* the animals that inhabit the region near the vent—oxidize these reduced substances using oxygen obtained from seawater. It is sometimes said that these vent communities are unique, in that they rely on geothermal rather than on solar energy. However, because each of the organisms in these communities ultimately depends on dissolved oxygen to oxidize reduced materials, and as that oxygen must come from green plant photosynthesis in terrestrial or shallow marine environments, it is clear that the deep-sea "geothermal oases" would shut down if photosynthesis at the surface of the earth were somehow stopped.

It is the "gutless" animals of the vent communities that are most relevant to the problem of the nature of the Ediacara fauna. *Riftia,* for example, is a cylindrical worm, up to 1.5 m in length and 4 cm in diameter, which, like other members of its phylum, lacks a mouth, digestive tract, and anus. Instead, there is a capacious internal organ called the trophosome (Greek., food body), the cells of which are packed sulfide-oxidizing bacteria. The H_2S and O_2 fuels are collected from the vent waters by filaments at the head of the worm, carried to the trophosome by the blood, and then used by the bacteria within the trophosome to provide the energy required for the synthesis of organic molecules. *Riftia* then consumes organic compounds supplied by the bacteria (Childress et al., 1987).

It is pretty clear that the Ediacara faunas were not vent communities. However, chemosymbiosis between animals and microbes is not limited to the neighborhood of hydrothermal vents; it is also found in other environments where the necessary ingredients (H_2S or CH_4, and O_2) occur in close proximity (Seilacher, 1990). For example, methane leaking from a sinkhole above a rising salt dome nourishes a sizable bed of deep-water mussels in the Gulf of Mexico (MacDonald et al., 1990). It may also be possible to identify this kind of activity in the fossil record; Savrda and Bottjer (1987) speculated that the Miocene bivalve *Anadara montereyana*, which lived at the interface between oxygen-containing (oxic) and oxygen-free (anoxic) waters, depended on chemosymbiotic microbes.

If the Ediacaran organisms contained chemosymbiotic bacteria, they must have lived at an interface between oxic and anoxic conditions. Such a boundary could have occurred at the sediment-water contact in sandy environments if a cyanobacterial mat had sealed the sediment surface, and if the oxygen content of normal seawater had been less than it is today (Runnegar, 1982b). Under such conditions, it is possible to envisage prostrate creatures such as *Phyllozoon* (Figure 3.10), *Dickinsonia* (Figure 3.13), and the "spindle-shaped organisms" from Newfoundland (Figure 3.7) as two-dimensional reaction chambers between oxic and anoxic microenvironments (Seilacher, 1989). Such a model may even apply to frondlike organisms such as *Charniodiscus* and *Rangea* (Figure 3.7); in these cases, the "fronds" may have extracted oxygen from seawater, and the "holdfasts" could have collected H_2S produced by bacterial sulfate reduction within the sediment. If this were true, it might be possible to interpret the rather thick "tentacles" that stud some "medusoids" as the "roots" of the "holdfasts." Even the knobbles of *Mawsonites* (Sun, 1986) could be interpreted in this way.

On the other hand, the upright, three-vaned "vendozoan," *Pteridinium* (Figure 3.7), does not fit so comfortably into this chemosymbiotic model because it seems to be all "leaves" and no "roots." Furthermore, *Pteridinium* has been found preserved in deep-water sediments in northern Canada (Narbonne and Aitken, 1990) and therefore is unlikely to have been photosymbiotic. *Pteridinium* has no body orifices and no structures for collecting particulate food. It seems possible that it was able to live on dissolved organic matter taken in through the body wall.

LEGLESS LOBOPODS, NAKED HALKIERIIDS, AND TOOTHED TERRORS

After some decades of study of both kinds of fossils, it is at last becoming possible to relate the Cambrian "small shelly fossils" to the animals preserved in the Cambrian *Lagerstätten*. The breakthrough began with *Wiwaxia corrugata* from the Burgess Shale (Figure 3.15), which Simon Conway Morris (1985b) showed to be a sluglike animal that was covered with blade-shaped plates called **sclerites.**

As preserved, the sclerites of *Wiwaxia* are unmineralized, but they may have been mineralized during life and have lost their minerals by solution after being buried. Mineralized *Wiwaxia*-like sclerites (made originally of aragonite) are common components of the small shelly faunas extracted from Early Cambrian limestones (Bengtson et al., 1990). They have been described under a number of names; the genus *Halkieria* is one that is both widespread and typical (Bengtson and Conway Morris, 1984). Some of these sclerites have an exceedingly complex internal structure (Bengtson et al., 1990; Jell, 1981).

It became clear to Bengtson and Conway Morris (1984) that *Halkieria* and its allies had been sclerite-bearing animals, just like *Wiwaxia*. They therefore published a model of *Halkieria* based on Conway Morris's reconstruction of *Wiwaxia*. Each halkieriid sclerite is much smaller than a *Wiwaxia* sclerite, so the model of *Halkieria* was very much like a scaled-down (no pun intended!) version of *Wiwaxia*. A few years later, articulated

FIGURE 3.15

Crushed specimen of *Wiwaxia corrugata* from the Burgess Shale (left), and a reconstructed dorsal view of the animal (right; from Dzik, 1986b, republished by permission). Photograph courtesy of S. Conway Morris.

 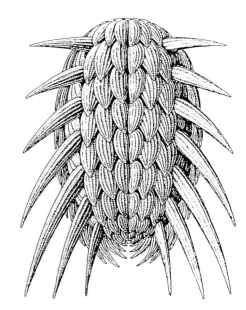

specimens of *Halkieria* were found in north Greenland (Conway Morris and Peel, 1990), and the reconstruction of *Halkieria* turned out to be wrong in two amazing ways: There were thousands of sclerites on *Halkieria*, rather than the 150 or so on *Wiwaxia*, and *Halkieria* had a limpet-shaped *shell* at each end of its body! In other respects (e.g., in the number of types of sclerite) the Bengtson-Conway Morris model proved to be correct.

Conway Morris (1985b), and Bengtson and Conway Morris (1984) regarded *Wiwaxia* and *Halkieria* as members of an extinct phylum close to the Mollusca. However, Butterfield (1990) has discovered evidence which, in his view, places *Wiwaxia* firmly within the Annelida (phyllodocid Polychaeta) rather than with *Halkieria* and its allies. The following sentence from Bengtson and Conway Morris (1984, p. 322) illustrates the nature of the problem: "Neither the general blade-shape, the longitudinal ribbing, nor the presence of a shaft-like attachment is any strong indication that *Halkieria* and *Wiwaxia* sclerites are homologous; all these characters are also found in pangolins" (scaly anteaters of Asia and Africa). Thus the halkieriids, ?*Wiwaxia,* and some of the other **cataphract** (sclerite-bearing) animals, represent either one, or possibly more, extinct clades of early Phanerozoic metazoans. The name now given to the halkieriids and their allies is the Coeloscleritophora (Bengtson et al., 1990); the Machaeridia may be a related or analogous clade (Dzik, 1986b).

A second breakthrough came with the discovery of the soft parts of an animal called *Microdictyon* (Greek., small net)—previously known only from its phosphatic netlike sclerites—in the Early Cambrian Chenjiang *Lagerstätte* (Bengtson et al., 1986; Chen et al., 1989). As reconstructed by Chen Jun-yuan, Hou Xian-guang, and Lu Hao-zhi, the 8-cm-long animal had a cylindrical body, 10 pairs of netlike sclerites arranged like the windows of an airplane along its flanks, and 10 pairs of appendages that might have been "soft legs." There are some similarities to another enigma from the Burgess Shale—appropriately named *Hallucigenia* by Conway Morris (1977)—but if *Microdictyon* really is a relative of *Hallucigenia,* then *Hallucigenia* will need to be turned on *its* head with the result that the long spines which Conway Morris regarded as legs become, in this new interpretation, weapons of defense.

But the *pièce de résistance* of the Cambrian *Lagerstätten* must surely be the champion of Stephen Jay Gould's book *Wonderful Life:* The monster of the Burgess Shale, *Anomalocaris* (Whittington and Briggs, 1985). Assembled from pieces that had been thought to be the parts of three or four different kinds of animals, *Anomalocaris* is now reconstructed as a sizable predator with grasping arms and a muscular mouth (Figure 3.16). It has been implicated in the origin of semicircular bites that are visible as healed injuries on the sides of Cambrian trilobites (Babcock and Robison, 1989), and it may well have been the first animal on earth to have evolved jaws with teeth.

One striking feature of each of these recently understood early animals is that they were, or are likely to have been, segmented. Many were also armed with spines, as were most trilobites and other early arthropods. The spines are often long and narrow and therefore do not appear to have been very strong; presumably, they were designed to discourage predatory worms that might have swallowed their prey whole, rather than to thwart hunters that had both jaws and teeth. When teeth and jaws evolved, the **cataphract armor** (a chain mail-like suit of spiny sclerites) became obsolete, just as it did for medieval soldiers.

It is instructive to consider what *Halkieria* might look like, stripped of its shells and sclerites. Its naked body may well have not been very different from *Spriggina, Dickinsonia,* or some of the other "segmented" animals of the Ediacara fauna. Thus a working hypothesis for further exploration of this fascinating stage of animal evolution might be the following: Animals began to crawl about on the seafloor at a time when the only predators were small and soft. They did so by becoming large and flat so that muscular energy could be transmitted more effectively to the ground. However, this shape became impractical once animals with teeth evolved, because a tooth-covered proboscis could rip an unprotected soft body to ribbons. The solution was to cover up with an armor made either of overlapping mineral sclerites or another kind of hard exterior, or alternatively, to hide from predators by burrowing into sand or mud. At this time, it was possible to avoid being swallowed whole by having spiny projections that would stick in the sides of a predator's gut; these are the weapons of defense of early trilobites and other arthropods, *Hallucigenia,* and members of the Hyolitha, Machaeridia, and Coeloscleritophora. This phase of animal evolution lasted through much of the Early and Middle Cambrian. However, when predators became more efficient, the possession of larger, continuous shells rather than overlapping sclerites became a more effective means of survival, and the cataphract metazoans declined in numbers and ultimately went extinct. Geerat J. Vermeij (1987) called the competitive coevolution of predator and prey *escalation;* it is an apt term.

THE CAMBRIAN EXPLOSION

The sudden appearance of many different animal phyla in Early Cambrian strata (Figures 3.3 and 3.6) has led to the view that animals must have had a substantial but invisible Precambrian history (e.g., Glaessner, 1984). The alternative hypothesis—which is also the null hypothesis—is that animals appeared in the fossil record soon after they first evolved. Thus we are faced with a choice between an "explosion of fossils" and an "explosion of

FIGURE 3.16

Marianne Collins's drawing of the two known species of *Anomalocaris* (from Gould, 1989; republished with permission).

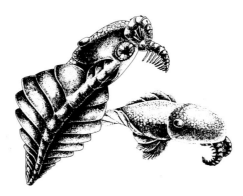

animals" to explain the nature of the Precambrian-Cambrian boundary (Runnegar, 1982b). Gould (1977) and Sepkoski (1978) favored the latter alternative, arguing that the abrupt appearance of many different kinds of animals at the beginning of the Cambrian represents no more than a log phase in the growth of animal diversity (Gould's "Sigmoid Fraud"). As with the modern increase in the human population, the log phase of a growth curve will seem to be instantaneous when viewed against a sufficiently long time scale.

These are two extreme alternatives. Either the late Precambrian-Early Cambrian fossil record should be read literally (null hypothesis) or it reflects the sudden appearance of fossilizable animals from forms that were unmineralized and too small or too flimsy to leave trace fossils. If the null hypothesis is correct, it becomes necessary to explain how so many different kinds of animals could have evolved almost instantaneously (Figure 3.6) and yet have apparently changed so little in the following 500 million years. Alternatively, if animals had a relatively long Precambrian history, why do so many kinds appear abruptly and more or less synchronously in the fossil record during Early Cambrian time? Was there an external environmental factor that either prevented animals from becoming fossilizable (i.e., large muscular, and/or mineralized) prior to the beginning of the Cambrian, or did some biological or environmental event trigger the metazoan radiation? Hypotheses that have been advanced to explain these questions tend to focus on either an external environmental control on, or an internal property of, the organisms that participated in the Cambrian explosion. Consequently, they are classified as either **extrinsic** or **intrinsic** hypotheses. Generally, intrinsic models have been used to explain the null hypothesis, and extrinsic explanations have been applied to the idea that animals have a long but invisible Precambrian history.

The only evidence for a sizable Precambrian history for the Metazoa comes, not from fossils, but from the proteins and nucleic acids of living organisms (Runnegar, 1982c, 1986; but see also Erwin, 1989). Let us look at a single example. As mentioned earlier, the relationships of living species can be assessed in a quantitative way by measuring the percentage similarity (or difference) between the sequences of the four kinds of nucleotides (A, G, C, or T) in the genes for molecules such as the ribosomal RNAs. Each possible pair of species may be compared in this way; it is then possible to calculate average differences between groups of species (Figure 3.17).

If representatives of some of the major living phyla are analyzed in this way, it can be shown that, on average, echinoderms and chordates are about 14% different in terms of their 18S rRNA sequences (Figure 3.17). Judged by this criterion, molluscs are a little more different from both chordates and echinoderms (~16%), but not to any significant extent. However, if cnidarians are compared with echinoderms, chordates, and molluscs, the observed differences are, on average, much greater—about 22% in each case. It seems, therefore, that as is expected from the comparative anatomy of these organisms, cnidarians left the line leading to chordates some time before the separation between echinoderms and chordates (Figure 3.17).

The echinoderm-chordate split can be no younger that about 520 Ma because fossil echinoderms are found in Early Cambrian rocks (Jefferies, 1990; Paul and Smith, 1984). So, if it has taken ~500 Ma for a 14% difference to develop between echinoderms and chordates, the split between cnidarians and all other Metazoa (Figure 3.1)—which resulted in a 22% difference (Figure 3.17)—must have occurred some time prior to 520 Ma ago. If the rate of evolution was approximately constant over long periods of time, the split between the Cnidaria and the other Metazoa could have been as early as 800 to 900 Ma before the present.

The Early Environment and the Rise of the Metazoa

In the last few years it has become increasingly obvious that the rise of the Metazoa—or at least their diversification into the rich fossil record of the early Phanerozoic—coincided

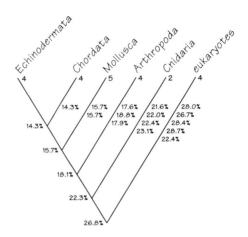

FIGURE 3.17

Phylogenetic tree (not drawn to scale) obtained from pairwise comparisons of 18S rRNA sequences from 19 species of animals (Field et al., 1988) and 4 species of other eukaryotes (ciliate, yeast, slime mold, and maize). The numbers at the tips of the branches are the number of species sampled in each branch ("Mollusca" includes the sipunculan *Golfinga*). The percentages are averages of the observed differences between pairs of species along all possible pathways in the tree. For example: eukaryote 18S rRNAs, on average, differ from echinoderm 18S rRNAs by 28.0%, from chordate 18S rRNAs by 26.7%, from mollusc and sipunculan 18S rRNAs by 28.4%, from arthropod 18S rRNAs by 28.7%, from cnidarian 18S rRNAs by 22.4%, and from the 18S rRNAs of all the animals in the tree by 26.8%. These data strongly suggest that the cnidarians separated from the other Metazoa long before the echinoderm-chordate split. See text for further explanation; the matrix of pairwise sequence differences was provided by James A. Lake.

approximately with a number of other exceptional worldwide events. For example, the end of the Precambrian is marked, not only by the near synchronous appearance in the fossil record of most of the living and extinct animal phyla and the earliest trace fossils, but also by the first unicelled organisms with mineral skeletons, new diverse assemblages of organic-walled microfossils called acritarchs (which, as discussed in Chapter 2, may be in part the resting stages of planktonic algae), the oldest known microbial endoliths, and many new kinds of megascopic calcareous algae. At the same time, the earth was recovering from the only ice age that has put glaciers at sea level in equatorial latitudes and a protracted period of deposition of marine carbonates that are abnormally enriched in the heavy isotope of carbon. These events also overlap the first global episode of sedimentary phosphate deposition, and they occur at a time when the amount of oxygen in the atmosphere and hydrosphere may have increased significantly. They also coincide with a possible secular change in the mineralogy of nonskeletal, sedimentary carbonates, and with the breakup of a long-lived and largely low-latitude supercontinent.

Is there any connection between these apparently unrelated biotic and environmental events? Molecular evidence from living organisms indicates that the animal, algal, and fungal kingdoms had a lengthy, but largely invisible, Precambrian history. The abrupt appearance of many different kinds of hard-bodied and soft-bodied fossils at the close of the Precambrian is therefore a result of the rise of large, resistant or mineralized organisms from tiny inconspicuous and insubstantial ones. In making this transition, it seems possible that small but unidirectional environmental changes were amplified by the global biota in an opportunistic fashion. In particular, early metazoans may have increased the oxygen content of the atmosphere by accelerating the accumulation of photosynthetically fixed carbon (via deposition of their fecal pellets) in oceanic environments. The effects of this increased rate of carbon burial may be reflected in the heavier carbonate carbon isotopes, the low-latitude glaciations (resulting from the "drawdown" of atmospheric CO_2 and a reversed "greenhouse" effect), the mineralogy of nonskeletal carbonates, and in the near synchronous appearance of calcareous skeletons in many unrelated groups of organisms, including protists, algae, and animals. The challenge is to find the evidence which can be used to develop quantitative models that will simulate the order and magnitude of the observed environmental and biological changes.

■

REFERENCES CITED

Aldridge, R. J., and Briggs, D. E. G. 1986. Conodonts. In A. Hoffman and M. H.Nitecki (Eds.), *Problematic Fossil Taxa* (New York: Oxford Univ. Press), pp. 227–239.

Anderson, M. M., and Conway Morris, S. 1982. A review, with descriptions of four unusual forms, of the soft-bodied fauna of the Conception and St. John's Groups (late Precambrian), Avalon Peninsula, Newfoundland. *Proc. 3rd N. Amer. Paleont. Convention 1:* 1–8.

Arai, M. N., and McGugan, A. 1968. A problematical coelenterate(?) from the Lower Cambrian, new Moraine Lake, Banff area, Alberta. *J. Paleont. 42:* 205–209.

Ax, P. 1989. Basic phylogenetic systematization of the Metazoa. In B. Fernholm, K. Bremer, and H. Jörnvall (Eds.), *The Hierarchy of Life* (Amsterdam: Excerpta Medica), pp. 229–245.

Babcock, L. E., and Feldmann, R. M. 1986. The phylum Conulariida. In A. Hoffman and M. H. Nitecki (Eds.), *Problematic Fossil Taxa* (New York: Oxford Univ. Press), pp. 135–147.

Babcock, L. E., and Robison, R. A. 1989. Preferences of Palaeozoic predators. *Nature 337:* 695–696.

Barthel, K. W. 1978. *Solnhofen. Ein Blick in die Erdgeschichte* (Basel: Ott Verlag Thun), 369 pp.

Bengtson, S. 1986. Introduction: The problem of the Problematica. In A. Hoffman and M. H. Nitecki (Eds.), *Problematic Fossil Taxa* (New York: Oxford Univ. Press), pp. 3–11.

Bengtson, S., and Conway Morris, S. 1984. A comparative study of Lower Cambrian *Halkieria* and Middle Cambrian *Wiwaxia. Lethaia 17:* 307–329.

Bengtson, S., Conway Morris, S., Cooper, B. J., Jell, P. A., and Runnegar, B. 1990. Early Cambrian fossils from South Australia. *Mem. Assoc. Austral. Palaeont. 9,* 364 pp.

Bengtson, S., Matthews, S. C., and Missarzhevsky, V. V. 1986. The Cambrian net-like fossil *Microdictyon.* In A. Hoffman and M. H. Nitecki (Eds.), *Problematic Fossil Taxa* (New York: Oxford Univ. Press), pp. 97–115.

Bergström, J. 1989. The origin of animal phyla and the new phylum Procoelomata. *Lethaia 22:* 259–269.

Bergström, J. 1990. Precambrian trace fossils and the rise of bilaterian animals. *Ichnos 1:* 3–13.

Butterfield, N. J. 1990. A reassessment of the enigmatic Burgess Shale fossil *Wiwaxia corrugata* (Matthew) and its relationship to the polychaete *Canadia spinosa* Walcott. *Paleobiology 16:* 287–303.

Chen Jun-yuan, Bergström, J., Lindström, M., and Hou Xian-guang. 1991. The Chengjian fauna—oldest soft-bodied fauna on earth. *National Geog. Res. Explor.* (in press).

Chen Jun-yuan, Hou Xian-guang, and Lu Hao-zhi. 1989. Early Cambrian netted scale-bearing worm-like sea animal. *Acta Palaeont. Sinica 28:* 1–16 (in Chinese with English abstract).

Childress, J. J., Felbeck, H., and Somero, G. N. 1987. Symbiosis in the deep sea. *Sci. Amer. 256*(5):107–112.

Clark, R. B. 1964. *Dynamics in Metazoan Evolution. The Origin of the Coelom and Segments* (Oxford: Clarendon Press), 313 pp.

Coates, M. I., and Clack, J. A. 1990. Polydactyly in the earliest known tetrapod limbs. *Nature 347:*66–69.

Collins, D., Briggs, D., and Conway Morris, S. 1983. New Burgess Shale fossil sites reveal Middle Cambrian faunal complex. *Science 222:* 163–167.

Conway Morris, S. 1977. Fossil priapulid worms. *Spec. Pap. Palaeontol. 20:* 95 pp.

Conway Morris, S. 1985a. Cambrian *Lagerstätten:* Their distribution and significance. *Phil. Trans. R. Soc. Lond. B311:* 49–65.

Conway Morris, S. 1985b. The Middle Cambrian metazoan *Wiwaxia corrugata* (Matthew) from the Burgess Shale and *Orygopsis* Shale, British Columbia, Canada. *Phil. Trans. R. Soc. Lond. B307:* 507–586.

Conway Morris, S., and Peel, J. S. 1990. Articulated halkieriids from the Lower Cambrian of north Greenland. *Nature 345:* 802–805.

Conway Morris, S., and Robison, R. A. 1986. Middle Cambrian priapulids and other soft-bodied fossils from Utah and Spain. *Univ. Kansas Paleont. Contrib. 117:* 1–22.

Corliss, J. B., and Ballard, R. D. 1977. Oases of life in the cold abyss. *Nat. Geogr. 152:* 441–453.

Craske, A. J., and Jefferies, R. P. S. 1989. A new mitrate from the Upper Ordovician of Norway, and a new approach to subdividing a plesion. *Palaeontology 32:* 69–99.

Devereaux, R., Loeblich, A. R., and Fox, G. E. 1990. Higher plant origins and the phylogeny of green algae. *J. Molec. Evol. 31:* 18–24.

Dzik, J. 1986a. Chordate affinities of the conodonts. In A. Hoffman and M. H. Nitecki (Eds.), *Problematic Fossil Taxa* (New York: Oxford Univ. Press), pp. 240–254.

Dzik, J. 1986b. Turrilepadida and other Machaeridia. In A. Hoffman and M. H. Nitecki (Eds.), *Problematic Fossil Taxa* (New York: Oxford Univ. Press), pp. 117–134.

Dzik, J., and Krumbiegel, G. 1989. The oldest 'onychophoran' *Xenusion:* A link connecting phyla? *Lethaia 22:* 169–181.

Emerson, M. J., and Schram, F. R. 1990. The origin of crustacean biramous appendages and the evolution of the Arthropoda. *Science 250:* 667–669.

Erwin, D. H. 1989. Molecular clocks, molecular phylogenetics and the origin of phyla. *Lethaia 22:* 251–257.

Exposito, J., and Garrone, R. 1990. Characterization of a fibrillar collagen gene in sponges reveals the early evolutionary appearance of two collagen gene families. *Proc. Nat. Acad. Sci. USA 87:* 6670–6673.

Fedonkin, M. A. 1984. Promorfologiya vendckikh Radialia [Promorphology of the Vendian Radialia]. *Stratigrafiya i Paleontologiya Drevneyshego Fanerozoya* (Moscow: Nauka), pp. 30–58.

Fedonkin, M. A. 1985a. Besskeletnaya fauna venda: Promorfologicheskiy analiz [Non-skeletal fauna of the Vendian: Promorphological analysis]. In B. S. Sokolov and A. B. Ivanovsky (Eds.), *Vendskaya Sistema. 1 Paleontologiya [The Vendian System. Vol. 1. Paleontology]* (Moscow: Nauka), pp. 10–69 (English translation: Berlin, Springer-Verlag, 1990, 7–70).

Fedonkin, M. A. 1985b. Precambrian metazoans: The problems of preservation, systematics and evolution. *Phil. Trans. R. Soc. Lond. B311:* 27–45.

Fedonkin, M. A. 1987. Besskeletnaya fauna venda i ee mesto v evolyutsii metazoa [The non-skeletal fauna of the Vendian and its place in the evolution of the Metazoa]. *Trudy Paleontol. Inst., Akad. Nauk SSSR 226:* 1–174.

Field, K. G., Olsen, G. J., Lane, D. J., Giovannoni, S. J. Ghiselin, M. T., Raff, E. C., Pace, N. R., and Raff, R. A. 1988. Molecular phylogeny of the animal kingdom. *Science 239:* 748–753.

Gehling, J. G. 1988. A cnidarian of actinian-grade from the Ediacaran Pound Subgroup, South Australia. *Alcheringa 12:* 299–314.

Glaessner, M. F. 1959. Precambrian Coelenterata from Australia, Africa and England. *Nature 183:* 1472–1473.

Glaessner, M. F. 1961. Pre-Cambrian animals. *Sci. Am. 204*(3):72–78.

Glaessner, M. F. 1969. Trace fossils from the Precambrian and basal Cambrian. *Lethaia 2:* 369–393.

Glaessner, M. F. 1984. *The Dawn of Animal Life* (Cambridge: Cambridge Univ. Press), 244 pp.

Glaessner, M. F., and Wade, M. 1966. The late Precambrian fossils from Ediacara, South Australia. *Palaeontology 9:* 599–628.

Goreau, T. F., Goreau, N. L., and Yonge, C. M. 1971. Reef corals: Autotrophs or heterotrophs? *Biol. Bull. 141:* 247–260.

Gould, S. J. 1977. *Ever Since Darwin. Reflections in Natural History* (New York: Norton), 285 pp.

Gould, S. J. 1989. *Wonderful Life. The Burgess Shale and the Nature of History* (New York: Norton), 437 pp.

Guoy, M., and Li, W. 1989. Molecular phylogeny of the kingdoms Animalia, Plantae and Fungi. *Molec. Biol. Evol. 6:* 109–122.

Hahn, G., and Pflug, H. D. 1988. Zweischalige Organismen aus dem Jung-Präkambrium (Vendium) von Namibia (SW-Afrika). *Geol. Palaeontol. 22:* 1–19.

Hill, D. 1972. *Treatise on Invertebrate Paleontology. Part E. Vol. 1. Archaeocyatha* (Boulder, CO: Geological Society of America), 158 pp.

Hou Xianguang, and Sun Weiguo. 1988. Discovery of Chengjiang fauna at Meishucun, Jinning, Yunnan. *Acta Palaeontol. Sinica 27:* 1–12 (in Chinese with English abstract).

Jacobs, D. K. 1990. Selector genes and the Cambrian radiation of Bilateria. *Proc. Nat. Acad. Sci. USA 87:* 4406–4410.

Jefferies, R. J. P. 1990. The solute *Dendrocystoides scoticus* from the Upper Ordovician of Scotland and the ancestry of chordates and echinoderms. *Palaeontology 33:* 631–679.

Jell, P. A. 1981. *Thambetolepis delicata* gen. et sp. nov., an enigmatic fossil from the Early Cambrian of South Australia. *Alcheringa 5:* 85–93.

Jenkins, R. J. F. 1985. The enigmatic Ediacaran (late Precambrian) genus *Rangea* and related forms. *Paleobiology 11:*336–355.

Jenkins, R. J. F., and Gehling, J. G. 1978. A review of the frond-like fossils of the Ediacara assemblage. *Rec. S. Austral. Mus. 17:* 347–359.

Kauffman, E. G., and Steidtmann, J. R. 1981. Are these the oldest metazoan trace fossils? *J. Paleont. 55:* 923–947.

Kirschvink, J. L. 1992. A paleogeographic model for Vendian and Cambrian time. In J. W. Schopf and C. Klein (Eds.), *The Proterozoic Biosphere, A Multidisciplinary Study* (New York: Cambridge Univ. Press), Chapter 12 (in press).

Lake, J. A. 1990. Origin of the Metazoa. *Proc. Nat. Acad. Sci. USA 87:* 763–766.

Landing, E., Narbonne, G. M., Benus, A. P., and Anderson, M. M. 1988. Faunas and depositional environments of the Upper Precambrian through Lower Cambrian, southeastern Newfoundland. *New York State Mus. Geol. Surv. Bull. 463:* 18–52.

Lewis, E. B. 1978. A gene complex controlling segmentation in *Drosophila. Nature 276:* 565–573.

Li, W., and Graur, D. 1990. *Fundamentals of Molecular Evolution* (Sunderland, MA: Sinauer), 284 pp.

Littler, M. M., Littler, D. S., Blair, S. M., and Norris, J. N. 1985. Deepest known plant life discovered on an uncharted seamount. *Science 227:* 57–59.

MacDonald, I. R., Reilly, J. F., Guinasso, N. L., Brooks, J. M., Carney, R. S., Bryant, W. A., and Bright, T. J. 1990. Chemosynthetic mussels at a brine-filled pockmark in the northern Gulf of Mexico. *Science 248:* 1096–1099.

Manhart, J. R., and Palmer, J. D. 1990. The gain of two chloroplasts marks the green algal ancestors of land plants. *Nature 345:* 268–270.

McMenamin, M. A. S. 1986. The garden of Ediacara. *Palaios 1:* 178–182.

Misra, S. B. 1969. Late Precambrian(?) fossils from southeastern Newfoundland. *Geol. Soc. Amer. Bull. 80:* 2133–2140.

Müller, K. J., and Walossek, D. 1985. Skaracarida, a new order of Crustacea from the Upper Cambrian of Västergötland, Sweden. *Fossils and Strata 17,* 65 pp.

Müller, K. J., and Walossek, D. 1987. Morphology, ontogeny and life habit of *Agnostus pisiformis* from the Upper Cambrian of Sweden. *Fossils and Strata 19,* 124 pp.

Müller, K. J., and Walossek, D. 1988. External morphology and larval development of the Upper Cambrian maxillipod *Brediocaris admirabilis. Fossils and Strata 23,* 70 pp.

Narbonne, G. M., and Aitken, J. D. 1990. Ediacaran fossils from the Sekwi Brook area, Mackenzie Mountains, northwestern Canada. *Palaeontology 33:* 945–980.

Narbonne, G. M., and Hoffman, H. 1987. Ediacaran biota of the Wernecke Mountains, Yukon, Canada. *Palaeontology 30:* 647–676.

Nitecki, M. H. (Ed.). 1979. *Mazon Creek Fossils* (New York: Academic Press), 581 pp.

Orlowski, S., and Radwanski, A. 1986. Middle Devonian sea-anemone burrows, *Alpertia santacrucensis* ichnogen. et ichnosp. n., from the Holy Cross Mountains. *Acta Geol. Polonica 36:* 233–249.

Patterson, C. 1989. Phylogenetic relations of major groups: Conclusions and prospects. In B. Fernholm, K. Bremer, and H. Jörnvall (Eds.), *The Heirarchy of Life. Molecules and Morphology in Molecular Analysis* (Amsterdam: Excerpta Medica), pp. 471–488.

Patterson, C. 1990. Reassessing relationships. *Nature 344:* 199–200.

Paul, C. R. C., and Smith, A. B. 1984. The early radiation and phylogeny of echinoderms. *Biol. Rev. 59:* 443–481.

Perasso, R., Baroin, A., Qu, L. H., Bachellerie, J. P., and Adoutte, A. 1989. Origin of the algae. *Nature 339:* 142–144.

Pflug, H. D. 1970a. Zur Fauna der Nama-Schichten in Südwest-Africa. I. Pteridinia, Bau and systematische Zugehörigkeit. *Palaeontographica A135:* 198–231.

Pflug, H. D. 1970b. Zur Fauna der Nama-Schichten in Südwest-Africa. II. Rangeidae, Bau and systematische Zugehörigkeit. *Palaeontographica A134:* 226–262.

Pflug, H. D. 1972a. Zur Fauna der Nama-Schichten in Südwest-Africa. III. Erniettomorpha, Bau und systematik. *Palaeontographica A139:* 134–170.

Pflug, H. D. 1972b. The Phanerozoic-Cryptozoic boundary and the origin of the Metazoa. *Proc. 24th Internatl. Geol. Congress, Montreal, Section 1:* 58–67.

Raup, D. M., and Seilacher, A. 1969. Fossil foraging behavior: Computer simulation. *Science 166:* 994–995.

Rice, M. E. 1985. Sipuncula: Developmental evidence for phylogenetic inference. In S. Conway Morris, J. D. George, R. Gibson, and H. M. Platt (Eds.), *The Origins and Relationships of Lower Invertebrates. Systematics Association Special Volume No. 28* (Oxford: Clarendon Press), pp. 274–296.

Runnegar, B. 1982a. Oxygen requirements, biology and phylogenetic significance of the late Precambrian worm *Dickinsonia,* and the evolution of the burrowing habit. *Alcheringa 6:* 223–239.

Runnegar, B. 1982b. The Cambrian explosion: Animals of fossils? *J. Geol. Soc. Australia 29:* 395–411.

Runnegar, B. 1982c. A molecular-clock date for the origin of the animal phyla. *Lethaia 15:* 199–205.

Runnegar, B. 1986. Molecular palaeontology. *Palaeontology 29:* 1–24.

Runnegar, B., and Pojeta, J. 1985. Origin and diversification of the Mollusca. In E. R. Trueman and M. R. Clarke (Eds.), *The Mollusca. Vol. 10* (Orlando, FL: Academic Press), pp. 1–57.

Runnegar, B., Pojeta, J., Morris, N. J., Taylor, J. D., Taylor, M. E., and McClung, G. 1975. Biology of the Hyolitha. *Lethaia 8:*181–191.

Saint, R. 1990. Homeobox genes in morphogenesis and tissue pattern formation. *Today's Life Science 2*(10): 14–21.

Savrda, C. E., and Bottjer, D. J. 1987. The exaerobic zone, a new oxygen-deficient marine biofacies. *Nature 327:* 54–56.

Seilacher, A. 1989. Vendozoa: Organismic construction in the Proterozoic biosphere. *Lethaia 22:* 229–239.

Seilacher, A. 1990. Aberrations in bivalve evolution related to photo- and chemosymbiosis. *Histor. Biol. 3:* 289–311.

Seilacher, A., Reif, W., and Westphal, F. 1985. Sedimentological, ecological and temporal patterns of fossil Lagerstätten. *Phil. Trans. R. Soc. Lond. B311:* 5–23.

Sepkoski, J. J. 1978. A kinetic model of Phanerozoic taxonomic diversity I. Analysis of marine orders. *Paleobiology 4:* 223–251.

Smith, T. L. 1989. Disparate evolution of yeasts and filamentous fungi indicated by phylogenetic analysis of glyceraldehyde-3-phosphate dehydrogenase genes. *Proc. Nat. Acad. Sci. USA 86:* 7063–7066.

Sprigg, R. C. 1947. Early Cambrian (?) jellyfishes from the Flinders Ranges, South Australia. *Trans. R. Soc. S. Austral. 71:* 212–224.

Sprigg, R. C. 1949. Early Cambrian "jellyfishes" of Ediacara, South Australia and Mount John, Kimberley District, Western Australia. *Trans. R. Soc. S. Austral. 73:* 72–99.

Sprigg, R. C. 1988. On the 1946 discovery of the Precambrian Ediacaran fossil fauna in South Australia. *Earth Sci. Hist. 7:* 46–51.

Sun Weiguo. 1986. Late Precambrian scyphozoan medusa *Mawsonites randellensis* sp. nov. and its significance in the Ediacara metazoan assemblage, South Australia. *Precamb. Res. 31:* 325–360.

Valentine, J. W. 1989. Bilaterians of the Precambrian-Cambrian transition and the annelid-arthropod relationship. *Proc. Nat. Acad. Sci. USA 86:* 2272–2275.

Vermeij, G. J. 1987. *Evolution and Escalation. An Ecological History of Life* (Princeton, NJ: Princeton Univ. Press), 527 pp.

Wade, M. 1968. Preservation of soft-bodied animals in Precambrian sandstones at Ediacara, South Australia. *Lethaia 1:* 238–267.

Wade, M. 1972a. Hydrozoa and Scyphozoa and other medusoids from the Precambrian Ediacara fauna, South Australia. *Palaeontology 15:* 197–225.

Wade, M. 1972b. *Dickinsonia:* Polychaete worms from the late Precambrian Ediacara fauna, South Australia. *Rec. Qld. Mus. 16:* 171–190.

Walossek, D., and Müller, K. J. 1990. Upper Cambrian stem-lineage crustaceans and their bearing upon the monophyletic origin of Crustacea and the position of *Agnostus. Lethaia 23:* 409–427.

Whittington, H. B. 1985. *The Burgess Shale* (New Haven: Yale Univ. Press) 151 pp.

Whittington, H. B., and Briggs, D. E. G. 1985. The largest Cambrian animal, *Anomalocaris,* Burgess Shale, British Columbia. *Phil. Trans. R. Soc. Lond. B309:* 569–609.

Wilkinson, D. G., Bhatt, S., Cook, M., Boncinelli, E., and Krumlauf, R. 1990. Segmental expression of Hox-2 homeobox-containing genes in the developing mouse hindbrain. *Nature 341:* 405–409.

CHAPTER 4

ORIGIN AND EVOLUTION
OF THE EARLIEST
LAND PLANTS

■

John B. Richardson†

> With regard to the origin of the land flora, . . . "the primary source of evidence has been the study of living Archegoniatae [higher plants]; but the reasoning based on these has been checked by palaeontological fact. This mode of enquiry should lead to more valuable results in the organographic study of Land-Plants at large than any mere search for phyletic schemes."
>
> F. O. Bower, *Primitive Land Plants* (1935; p. 635)

■

INTRODUCTION

The ancestors of higher land plants, progenitors of the liverworts, hornworts, and mosses **(bryophytes)** and of the ferns and seed plants **(tracheophytes)** of the present-day flora, successfully invaded the land in the early Paleozoic. Rocks deposited about 460 Ma ago (during Caradoc times of the Ordovician Period) contain irrefutable evidence for their existence. The nature of these pioneers, their habitat and life cycles, and their impact on the terrestrial environments are, however, subjects of conjecture and much debate. Terrestrial environments, at least freshwater lakes and river systems, and possibly also fine silts deposited during floods, rock waste from screes, and even some desert settings, must have been inhabited by photoautotrophs in Precambrian times. The earliest land colonizers were probably cyanobacteria, organisms surviving today alongside a land vegetation that includes terrestrial algae, but that is usually dominated by higher plants **(embryophytes)**, the liverworts, hornworts, mosses, and tracheophytes.

How did these higher plants evolve and what are their interrelationships? Two main approaches have been used to study their evolutionary patterns. First, based on studies of living plants, botanists have attempted to classify members of the modern flora, deduce their relationships, and to reconstruct their **phylogeny,** or group history. Second, based on

†Department of Palaeontology, British Museum (Natural History), Cromwell Road, SW7 5BD, London, England

studies of the morphology of fossil plants and their spores, palynologists and paleobotanists have carefully documented the time of first appearance and the distribution of various plant types through time and used these data to reconstruct phylogenetic relationships. Some proponents of the first method ignore the fossil record entirely. Instead, they rely exclusively on shared and evolutionarily derived characters among living plants, and the rules of **parsimony** ("the simplest solution being most likely to be correct") to reconstruct phylogenies, only considering information from living plants. In this method, a series of characters is studied, and a decision is made as to which of these characters are "primitive" and which are "derived"; from this information, the plant phylogeny is reconstructed.

Mishler and Churchill (1984, 1985) have made rigorous recent attempts at analyses of this type, termed **cladistic analyses,** applying them to living spore-producing (**crypto-gammic**) plants and, especially, to bryophytes. The cladistic method is excellent for analyzing possible relationships, but it tells us nothing about the timing of the evolutionary events leading to these relationships, events resulting in sequential phases of greening of the landscape, or about the nature and distribution of early, now extinct, higher plants. The only direct evidence of past floras is fossil evidence, but this is often fragmentary, imperfectly known, and biased toward preservable, resistant tissues. Thus, for example, undisputed **gametophytes** (that portion of the plant life cycle producing sex cells, or **gametes**) are rare, or unknown, in Ordovician, Silurian, and Lower Devonian strata, although it is possible that, for at least some of the plants of the Late Ordovician and Early Silurian, the gametophytic plant was the dominant portion of the life cycle as it is in modern bryophytes. Consequently, a combination of methods, botanical and paleobotanical-palynological, must be used to achieve progress toward understanding the nature of early land floras.

The Lyellian axiom "the present is the key to the past" reminds us that the basic principles of biology, chemistry, and physics, and therefore of geology, have remained unchanged through time. But in trying to understand ancient life, it is essential to remember that the earth is dynamic and that, in the words of the novelist L. P. Hartley; "The past is another country; they do things differently there." Indeed, the plants now known from early Paleozoic strata are far different from those living today, having a floral diversity undreamed of a scant two decades ago. Certainly, the early Paleozoic landscape would fascinate a time-traveling botanist!

PROBLEMS ASSOCIATED WITH LIFE ON LAND

Living in water, rather than on land, has several obvious advantages. Water provides a supporting protective medium that contains dissolved nutrients and gases, is less subject to extremes of temperature than air, and that serves as a medium for transport of motile (including sexual) phases of life cycles. In contrast, life on land is subject to water stress, greater seasonal and diurnal temperature change, and to fluctuations in UV radiation. In consequence, land plants have evolved structures and strategies to combat these environmental factors. Combinations of these structures preserved in fossil plants tell us about the adaptation of plants to life on land and their evolutionary progression over time. Like early amphibians (see Chapter 5), however, the earliest land plants (free-sporing embryophytes) were dependent on water for transport of the motile sexual phases of their life cycles.

Land Plant Structures

Adaptations to living in a subaerial environment are reflected in a number of structures occurring in fossil plants. Such features, providing clues to the nature of extinct plant groups and their mode of life, are as follows:

1. *Spores:* Small (usually microscopic), aerially distributed, reproductive bodies having highly resistant coats (exines), which in modern spores are commonly composed of **sporopollenin** (an organic substance also found in some algae and algal cysts).

2. *Cuticle:* A waxy coating, occurring on the external surfaces of plants, that prevents water loss and possibly also protects the plant from pathogens and deleterious effects of UV radiation.

3. *Stomata:* Pores, occurring in plant stems and leaves, used to control water loss and gaseous exchange through the cuticle.

4. *Vascular tissue:* Water- and nutrient-conducting tissue, axial tubelike strands consisting of thick-walled **xylem** (chiefly for water conduction) and thin-walled **phloem** (chiefly for nutrient conduction); the xylem may be **lignified** (thickened and strengthened by lignin, a type of organic compound) and possess internal thickenings. Mishler and Churchill (1984) divide vascular tissue into two main types: (a) hydroids and leptoids, xylem- and phloemlike structures that occur in mosses but which perhaps contain "protolignin" rather than lignin; and (b) true xylem and phloem, conducting cells in which the xylem can be highly lignified with "true lignin," providing support for the erect plant (the **sporophyte,** the spore-producing portion of the life cycle); the xylem constitutes the woody tissue, or tracheids.

5. *Water absorbing organs* and intercellular air spaces.

Strategies for Life on Land

A large amount of the available evidence for early land plants is from Late Silurian and Devonian rocks of the North Atlantic area, a region referred to as "Laurussia," or the "Old Red Sandstone Continent." Much of this continental mass lay within the tropics, and its climate has been interpreted as ranging from seasonally wet to that of a marginal desert. The nature of the preserved sediments suggests that large quantities of water occurred at times in some of the area. Early Devonian caliche deposits (mineral crusts that form today in arid regions), occurring in the Welsh Borderland, have been interpreted by Allen (1974) as horizons in fossil soils formed under semiarid conditions similar to those producing such mineral crusts today. Some recent caliches are estimated to have formed over a few to several thousand years in climatic regimes with mean annual temperatures of 12° to 20°C, and with a seasonally distributed rainfall of 100 to 500 mm. According to Allen, the rainfall is unlikely to have peaked seasonally because caliches "only grow where the monthly maximum is less than approximately 25% of mean annual precipitation" (Allen, 1974). Certain Early Devonian caliches ("calcrete" horizons) are massive, and one such horizon, which can be traced throughout much of the Anglo-Welsh Basin, may reflect the occurrence of a widespread, major climatic event. This event, however, was not entirely catastrophic, because at least one plant lineage, and several trends evidenced by the fossil spore flora, can be traced across it. Nevertheless, few of the distinctive spore species from the uppermost Silurian survived into the Devonian.

Berkner and Marshall (1965) postulated that the Siluro-Devonian atmosphere contained markedly less oxygen, and therefore provided a less effective UV-absorbing ozone screen, than that of the present. However, more recent estimates suggest that the atmosphere of this time contained appreciable amounts of oxygen, between 20% and 50% of the present atmospheric level (Raven, 1985), and recent research shows that at least some plants can adapt to fluctuations in UV radiation.

How do higher plants cope with the stressful land environment, and why did they invade at all if this environment was so inhospitable? The answer may be that the earliest land plants evolved from photoautotrophs that were already on "land," living in freshwater rather than marine environments, and that the impetus for their evolution was at least in part climatic. Following a Late Ordovician ice age (see Figure 2.2), the areas where many of the fossil plants are found had, by Early Devonian time, become tropical. This spans a period of about 30 Ma. Bryophytes (for example, mosses) thrive in conditions that range from subglacial to tropical, and many species have become widely distributed since the last ice age, some 10,000 years ago (Schofield, 1985). In terms of geological time, dispersal within several to even many thousands of years is essentially "instantaneous." Such bryophytes and other land plants have a number of successful strategies for thriving in subaerial environments:

1. *Short life cycles.* Some small, ephemeral modern mosses have telescoped life cycles; these forms are prevalent only in the early spring or the late autumn and thus avoid more unfavorable conditions. In spite of the relatively large size of their spores, they are widely distributed (Crum, 1972). For early land plants, this capability for wide dispersal would have been important for survival in ephemeral environments and could have been combined with living in, or on, water, as do some liverworts (e.g., the sphaerocarpalean *Riella,* and species of the marchantialeans, *Riccia* and *Ricciocarpus*).

2. *Drought resistance.* This strategy involves the capability of tissues to dry out and to then rehydrate without loss of function. Some moss spores can germinate after being desiccated for as long as 16 years. Specimens of one moss species *(Tortula ruralis)* have been air-dried for ten months and shown to recover normal photosynthetic function within a few hours after rehydration (Schofield, 1985). Some liverworts have developed a different strategy for drought resistance: In *Riccia* (Marchantiales), plants of the spore-producing generation are protected within the tissue of the dead gamete-producing plant. In another group of liverworts (the Sphaerocarpales), the spores are often produced, and dispersed, in groups of four called tetrads. Each tetrad can be enclosed in an envelope that possibly further increases tolerance of desiccation. Many liverworts have life cycles in which both male and female gamete-producing plants are formed; that is, they are **dioecious.** In these liverworts, a spore tetrad can consist of two spores giving rise to male gamete-producing plants, and two spores giving rise to female egg-producing plants. Schofield (1985, p. 137) estimates that more than 90% of all liverworts are dioecious. Because the male gametes cannot travel farther than about 1 cm, "permanent" tetrads of this type may be an adaptation ensuring a high fertility rate in ephemeral habitats.

3. *Control of water loss by anatomical modifications.* The presence of a cuticle in which the stomata are associated with intercellular spaces is a feature that allows land plants both to regulate their rate of water loss and to maintain a water-conserving gas exchange system (Raven, 1977, 1985). Taylor (1982) considers that particular aspects of cellular organization of the charophyte algae, possible precursors of the earliest land plants, may be important in the early development of this system.

4. *Photoprotective adaptations.* Recent research has shown that certain plant pigments known as flavonoids have an important photoprotective function. These pigments are synthesized in plants when they are exposed to specific types of UV radiation, and their synthesis is sensitive to small changes in the amount of this radiation (Glasgow, 1990). Flavonoids are widespread among liverworts, mosses, hornworts, ferns, and seed plants (i.e., the embryophytes).

LIFE CYCLES AND SPORE TERMINOLOGY

A substantial portion of the following sections of this discussion deals with spore morphology and the life cycles of early evolving spore-producing plants. Before embarking on this discussion, therefore, a brief review of terminology is in order.

As we noted, free-sporing plants have two phases, or generations, in their life cycles: The spore-producing generation, the **sporophyte,** and the gamete- (sex cell-) producing phase, the **gametophyte.** In higher plants, spores are always produced by **meiosis,** sometimes called "reduction division," and the spores are therefore **haploid,** having one-half of the set of chromosomes that occurs in the body cells of the "adult" sporophyte plant.

Spores are produced in **sporangia** ("spore sacs"), and because they are products of meiosis they are produced in groups of four, commonly packed together in a pyramid-shaped **tetrahedral tetrad.** In such tetrads, the inner face of each spore, where it comes in contact with the other three spores of the tetrad, is termed the **proximal face,** whereas the outermost surface of each spore, the side of the spore that is on the outermost side of the pyramid, is the **distal face.** On their proximal faces, each of the spores is adpressed closely against the other three spores of the tetrad, making a three-rayed, more or less Y-shaped groove, or "suture," the so-called **trilete mark.** Distal faces, of course, lack trilete marks (and spores lacking grooves on all faces are termed **alete**).

The resistant coat, or wall, of a higher plant spore is termed the **exine.** In many taxa, the exine is composed of two more or less distinct layers, an inner layer, the **intexine,** and an outer smooth or sculptured layer, the **exoexine.** Spore size and shape, the structure of the exine layers and, particularly, the nature of the sculpture pattern expressed on the outer surface of the spore, are useful taxonomic characters. Spores produced by a particular plant are usually all of identical morphology, or fall within a limited, definable range of morphologies; the spores produced by different plant species are usually morphologically distinguishable. In fossil material, it is particularly useful to study spores that can be isolated from a sporangium attached to a plant axis, because the parent plants of such **in situ spores** can be directly identified. However, spores are produced in prodigious quantities, and, upon their maturation, are usually released from the sporangium in which they have been produced; such **dispersed spores** can be related to a parent plant if the same spore type has been found within a sporangium of known parentage.

After they are dispersed from the sporangia, some spores remain together in more or less permanent tetrads and may be held together by an encompassing envelope; in some taxa, these tetrads become subdivided into smaller two-spore packets **(dyads)**, or are further subdivided into their component single spores **(monads).** Tetrads, dyads, and monads of this type are termed **cryptospores.** In many spore-producing plants (ferns, for example), it is more common for individual spores to be released from the sporangia; dispersed single spores of this type that are of uncertain biological nature (i.e., for which there is no evidence whether their parent plants produced one type of spore, **isospores,** or produced specialized sexually differentiated spores, male **microspores** and female **megaspores**) are referred to as **miospores.**

During germination, spores produce a **germ tube,** which commonly emerges through the central portion of the trilete mark on the proximal face. Ultimately, the germinated haploid spores give rise (dividing by **mitosis,** or "body cell division") to small gamete-producing plants (the gametophyte generation). The **male gametes** (sperm) are produced in specialized portions of the gametophyte plant termed **antheridia,** whereas **female gametes** (eggs) are produced in **archegonia.** To carry out fertilization, the flagellated sperm must swim to the archegonia; water (a moist film is usually sufficient) is therefore needed for sexual reproduction. If both the antheridia and archegonia occur in a single gametophyte plant, the gametophyte is said to be **monoecious.** However, if separate male plants (gametophytes having antheridia) and female plants (gametophytes having archegonia) are produced, the gametophyte generation is said to be **dioecious.**

Because higher plant spores are haploid (products of meiosis), and because the gametophyte is produced via mitotic growth (in which each cell produced has the same number of chromosomes as its parent cell), the gametophyte generation is also haploid. During fertilization, fusion of the haploid gametes **(syngamy)** produces a diploid **zygote;** the zygote grows, via mitosis, to produce an embryo and, ultimately, the diploid, spore-producing (sporophyte) generation, completing the plant life cycle. Female gametes (eggs) are nonmotile; fertilization therefore normally occurs within an archegonium. For this reason, the sporophyte generation usually grows on top of, emerging out of, the archegonium of the gametophyte generation and, in some cases (e.g., mosses, hornworts, and liverworts), the sporophyte is **epiphytic** and remains dependent on the gametophyte.

CLADISTIC ANALYSIS OF THE RELATIONSHIPS OF SPORE-PRODUCING PLANTS

Before considering the fossil record of the earliest higher land plants, let us first examine the suggested relationships of spore-producing (cryptogammic) plants, based on the cladistic analysis carried out by Mishler and Churchill (1984, 1985).

Origin of Higher Land Plants

Evidence regarding the origin of bryophytes and tracheophytes—that is, the embryophytes—comes entirely from studies of living plants. The nature of photosynthetic pigments, of starch storage products, and of flagellar structure, all indicate that these higher plants were derived from the green algae. And there is a further series of characters indicating that among the green algae, the Charophyceae (charophytes) are closest to the embryophytes. According to Mishler and Churchill (1984), the following characters indicate closer affinity between the higher plants and charophytes than with any other green algal group: (1) The occurrence in the two groups of symmetrical sperm; (2) the presence in both groups of a phragmoplast (a structure involved in cell wall production during cell division), a feature known in only one other group of green algae in addition to the charophytes; (3) the shared occurrence of the enzyme glycolate oxylase; and (4) the occurrence of similar flavonoids.

Relationships Within the Embryophytes

The evolutionary source of tracheophytic plants may be within the bryophytes because these plants are often pioneers and because early land plants bear some resemblances to them. In most bryophytes, a thin cuticle is present, but little is known of the composition and distribution of this layer throughout the group (Mishler and Churchill, 1985). Among photoautotrophs, the nature of the egg-producing cells (archegonia), the sperm-producing cells (antheridia), and the embryo (resulting from fusion of egg and sperm) are probably unique features of the higher land plants; unfortunately, however, under normal circumstances, these structures are not preserved in fossils. Stomates occur in the hornworts, mosses, and tracheophytes (but not in liverworts), although in some moss sporophytes they may be nonfunctional and in others they are absent (Bold et al., 1980, p. 258). Production of spores over a prolonged period, rather than all at one time, is another feature common to bryophytes (with the exception of the hornworts) and tracheophytes. As we noted, occurrence of the vascular tissues (xylem and phloem) is a unique feature of tracheophytes, although Mishler and Churchill (1984) tentatively concluded that the conducting tissues of mosses (hydroids and leptoids) and those of tracheophytes (xylem and phloem) are **homologous,** inherited from a common ancestor that had a rather simple level of vascular tissue. According to this interpretation, the primitive vascular tissue has since been reduced, or lost, in several groups of mosses (for a possible explanation of this reduction, or loss, see Hébant, 1977).

Although mosses contain lignin-*like* compounds (Mishler and Churchill, 1984), there has been a prolonged debate as to whether they, like tracheophytes, actually possess lignin (reviewed by Hébant 1977, and by Markham and Porter, 1978), a question complicated by the difficulty of chemically defining this compound. Using a strict chemical definition, the consensus is that lignin is not present in bryophytes, although there is evidence that biochemically related compounds, known as polyphenolics, are found in the conducting tissue of some mosses. Thus the presence of polyphenolics ("protolignin") in mosses has been regarded as a primitive character in a transition leading to "true lignin."

The spore-producing sporophyte phase of the "primitive" land plant envisaged by Mishler and Churchill (1984) was small, epiphytic, (growing on the gamete-producing phase of the plant), and consisted only of a simple "spore sac," or sporangium. They considered that the presence of erect, persistent, differentiated tissue-containing, un-branched sporophyte axes is a character of the moss-tracheophyte lineage. Inasmuch as only tracheophytes developed branched axes and multiple sporangia, does this interpretation mean that the earliest branched axes had to have attained the tracheophyte level of organization? Bower (1935, p. 591 and Figure 440), comparing the architecture of *Rhynia*, an early occurring fossil tracheophyte, with that of living plants, notes that branching of sporophytes "has been recorded as an occasional abnormality in living Bryophytes [mosses] . . . and [that] instances [of branching] have also been found in liverworts." The earliest tracheophyte-like fossil axes now known in the geologic record are dichotomously branched (J. M. Schopf et al., 1966). Was this a character of moss-tracheophyte ancestors that was partly lost in the moss, but not in the tracheophyte, descendents?

Liverworts, hornworts, and those tracheophytes having free-living gametophytes (that is, those nonseed plants in which egg- and sperm-producing plants are free-living) have unicellular rootlike structures (**rhizoids**), whereas mosses have multicellular rhizoids. Few, if any, undisputed fossil gametophytes are known, but *Rhynia gywnne-vaughani*, thought to be a gametophyte by Lemoigne (1968), has nonseptate rhizoids (Gensel and Andrews, 1984, p. 64).

Three further characters suggested by Mishler and Churchill (1984) for the earliest tracheophytes are the following: (1) the occurrence of leaves on the gametophyte (although this life cycle phase would not be detected easily in the fossil record, leaves are not seen on *Lyonophyton*, a possible gametophyte from the Devonian Rhynie Chert; Remy and Remy, 1980); (2) independence of the sporophyte from the gametophyte generation (a feature difficult to detect in fragmentary compression fossils); and (3) the occurrence of branched sporophytes having multiple sporangia (for which there is ample evidence for increasing elaboration in Early Devonian tracheophytes).

Problems

The large number of shared, evolutionary derived characters among various groups of land plants means that relatively few unique primary characters that could be used to delineate their origin remain in some groups (e.g., *Sphagnum* and its relatives). Overall, there is a trend toward reduction or loss of characters, especially in mosses (for example, in their vascular tissue). However, the *Sphagnum*-type mosses (the Sphagnaceae) are one of the few groups of living mosses with trilete spores, a character regarded here as primitive (that is, **pleisiomorphic**).

Conclusions

Mishler and Churchill (1984) concluded that there is strong evidence that all embryophytes (except the hornworts) are derived from a common ancestor (that is, that the group is

monophyletic). They believe that the mosses and liverworts share only such "primitive" (plesiomorphic) characters as small size and dominance of the gametophytic phase of the life cycle. According to this view, hornworts are more enigmatic; the most parsimonious position is that they are a parallel group to the moss-tracheophyte lineage (but, see also Blackmore and Barnes, 1987). Thus the bryophytes are regarded as being **paraphyletic** (i.e., a group not including all descendants of a common ancestor) and, thus, a group that should not be classified formally together.

There is also the question of the evolutionary origin of the alternation of the gametophyte and sporophyte generations in embryophytic plants. In the bryophytes, the gametophyte is the dominant generation, whereas in the evolution of tracheophytes the sporophyte stage becomes progressively dominant. Derivation of embryophytes from charophyte algae would involve changes in the charophyte life cycle (specifically, the delaying of the time of meiosis). As Stewart and Mattox (1975) pointed out, the nature of the charophyte life cycle makes it likely that the multicellular sporophyte stage of the embryophytes was "interpolated" between meiosis (spore formation) and syngamy (fusion of sex cells) as Bower (1908) originally proposed.

Mishler and Churchill (1984) have predicted that the ancestral form for the moss-tracheophyte lineage was an erect, unbranched sporophyte having vascular tissue (but not containing tracheids) and cuticle with stomates, that was epiphytic on a leafless, platelike ("thalloid") or possibly erect ("axial") gametophyte. Interestingly, this is not so different from plant fossils already known in Late Silurian- and Devonian-age strata.

Archetypes

In Figure 4.1A, a **cladogram** (an "evolutionary tree" of the type prepared by cladists), showing a series of hypothetical stages in the evolution from a liverwort-like plant to a

FIGURE 4.1

Comparison of the evolutionary relationships among green algae and embryophytic plants based on the cladistic analysis (A, above) of Mishler and Churchill (1984), and of the known palynological-paleobotanical fossil record (B, below).

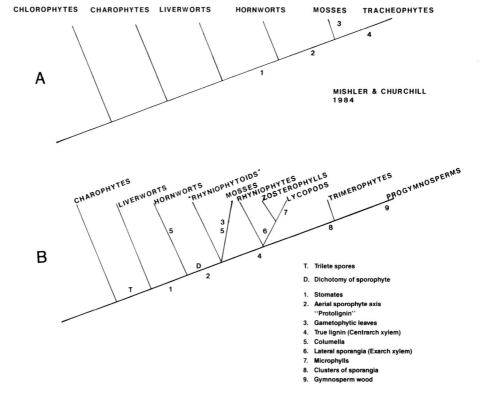

branching sporophytic tracheophyte (Mishler and Churchill, 1985, Figure 5), is compared with the known record of fossil plants (Figure 4.1B). In the following portions of this discussion, we compare available palynological and paleobotanical data in some detail with the four hypothetical stages, or **archetypes,** on which the cladogram is based. These four stages can be summarized as follows:

1. *The archetype for land plants:* A plant having a simple, flat, platelike ("thalloid") gametophyte generation and simple, attached "spore sacs" (sporangia).

2. *The archetype for hornworts, mosses, and tracheophytes:* A plant with a simple thalloid gametophyte generation, as for land plants, but having a sporophyte that bore stomata and that had an embedded "foot" portion and thrust-out "spore sacs" (morphological features present in modern hornworts).

3. *The archetype for mosses and tracheophytes:* A plant with a radially symmetrical, branched, "naked" gametophyte (lacking leaves), and an elevated, permanently epiphytic, stomata-bearing sporophyte having a conducting strand of xylem and phloem and a single sporangium (the hypothesized radial symmetry of the gametophyte being supported by the radial symmetry of the mature gametophytes of all living mosses).

4. *Sporophytic evolution within the tracheophytes* involved increases in plant size, branching, and in a number of sporangia, development of increasingly complex tracheids and, eventually, of leaves.

Although occasional windows on the past are opened by petrifactions, fossils showing exquisite anatomical detail (for example, those of the Rhynie Chert), because of their delicate nature it is unlikely that direct evidence will be found in the fossil record for primitive gametophytes of the type envisaged by Mishler and Churchill (particularly, for archetypes 1 and 2 as just described). On the other hand, painstaking work by Edwards and coworkers, on more commonly occurring fossil material preserved by compression, together with studies of fossil spores, are rapidly changing concepts of the nature of the Silurian and the earliest Devonian floras (Edwards and Fanning, 1985).

■

PALYNOLOGICAL-PALEOBOTANICAL ANALYSIS OF THE EARLY FOSSIL RECORD

We refer to the major innovations in plant and spore morphology and their periods of development as *evolutionary phases*. It should be emphasized that these phases represent stages in land plant evolution that are imperfectly understood. There is a growing wealth of data from studies of fossil spores (Gray, 1985; Richardson, 1985, 1988; Richardson and Ioannides, 1973; Richardson and Lister, 1969; Strother and Traverse, 1979), but until recently relatively little was known of the parent plants producing these spores (Edwards et al., 1979; Edwards and Fanning, 1985; Edwards and Feehan, 1980; Fanning et al., 1991; Pratt et al., 1978). Current evidence shows that whereas spore evolution was rapid, and the spore data may evidence the beginnings of several evolutionary lineages occurring in more advanced plants, the sporangia of such plants are more difficult to differentiate and may appear closely similar or identical to one another (Fanning et al., 1988). The evolutionary phases we discuss should not be confused with biostratigraphical zones, which are of much shorter duration than the phases. These five phases, encompassing the appearance of major plant innovations as reflected in the fossil record, are outlined for comparison with the archetypes suggested by Mishler and Churchill (1985, Figure 5) that we just listed. The following examples are based on fossils from the Laurussian paleocontinent, largely from the British Isles.

Evolutionary Phase 1: The Exclusive Cryptospore-Cuticle Phase of the Late Ordovician and Early Silurian (Caradoc and lower Llandovery)

Fossil assemblages in phase 1 consist solely of "permanent" spore tetrads and pairs (dyads), single alete spores (monads), sheets of cuticle, and isolated smooth tubes. The cuticles and spores may be remnants of different plants; no relationship has been established between specific spores and the cuticles. Gray (1985) compared the tetrads (known as cryptospore "permanent" tetrads) to the spores of some living liverworts and pointed out that spores closely similar in morphology to cryptospore tetrads and dyads occur in several liverwort orders. In the modern liverwort *Haplomitrium,* for example, the spores commonly have a tendency to adhere together in both tetrads and dyads; "permanent" tetrads of this type are also common in the liverwort *Sphaerocarpos,* in which an envelope may surround each tetrad. *Sphaerocarpos* has a spherical sporangium, with a small bulging basal "foot," similar to that suggested for archetype 1. Liverworts have several strategies that make them ideal pioneering plants. For example, they live on algal crusts on land, in shallow ephemeral pools, and in soils together with blue-green algae (cyanobacteria); and the nitrogen-fixing cyanobacterium *Nostoc* is able to live within the tissues of some liverworts and hornworts (that is, *Nostoc* is **endophytic** in these plants, providing them with biologically usable nitrogen). Early land pioneers, perhaps inhabiting fine mineral debris covered by cyanobacterial crusts, may well have found it advantageous to host a microbial nitrogen fixer of this type.

In trilete miospores, germination (production of the germ tube) is assumed to occur via the proximal sutures, but it is variable and in "permanent" tetrads and dyads germination may have been equatorial or even distal, as is known to occur in some modern liverworts (Inoue, 1960; Schofield, 1985). Schofield (1985, p. 201, Figure 15-5) illustrates a "permanent" tetrad of the liverwort *Sphaerocarpos stipitatus* in which each of the four germinating spores is producing its own germ tube. If germination of these spores were to produce both male and female gametophytes, as occurs in some modern liverworts in which there is sexual differentiation among the spores in a single tetrad, such germination would be advantageous, ensuring that the resulting male and female gametes are in close proximity to each other. No fossil plant axes are known from Late Ordovician-Early Silurian strata; thus the data from the plant fossil and spore record, sketchy though they are, resemble the archetypal model.

Using the Mishler and Churchill model (1985, Figure 1), the fossil cryptospores and sheets of cuticle present in the Late Ordovician and Early Silurian would fit either within the stem group of the liverwort-hornwort-moss-tracheophyte lineage or within the stem group of the latter three groups only, but apparently prior to the development of stomates. These early Paleozoic plants may have been ancestors of liverworts, hornworts, mosses, and tracheophytes, or of the latter three groups only; therefore, as Gray has suggested, these early plants appear to have been "liverwort-like."

In Late Ordovician and Early Silurian (Caradoc to early Llandovery) times, some land plants may have had flat, platelike ("thalloid") gametophytes and sporophytes lacking hairlike stalks (setae) that produced cryptospore tetrads, dyads, and alete monads that were commonly enclosed within either smooth or textured (sculptured) envelopes. The occurrence of tetrads, dyads, and monads having identical sculpture patterns may indicate that they were produced either by the same plant or by plants that were closely related. In some modern liverworts (specifically, those of the Sphaerocarpales and Marchantiales), spores are used in the differentiation of species. If tetrads, dyads, and monads (for example, those with identically sculptured envelopes) were produced by the same or related fossil plants, then there is the question as to what advantage this may have conferred. Because most modern hepatics (liverworts) are dioecious, having separate male and female gametophytes, bisexual tetrads and dyads would be a positive advantage at times when their habitats were both limited and ephemeral. The monads, spores of smaller size, may have promoted more rapid dispersal in more favorable conditions. There is a tendency for larger compound cryptospores (tetrads and dyads) to decrease in numbers in progressively younger geological units.

Evolutionary Phase 2: The Sporophyte-Trilete Isospores Phase of the Early to Middle Silurian (upper Llandovery and lower Wenlock)

This phase was characterized by the appearance in the fossil record of simple **dichotomizing** (bifurcating) plant axes (evidenced by the genus *Eohostinella*, possibly of Telychian age), and of banded, "vascular tissuelike" isolated tubes (late Llandovery and early Wenlock). Smaller, smooth-walled tubes, isolated or in clusters, persist from phase 1, and in younger strata are found intimately associated with banded tubes. The presence of fossil elevated sporophytes indicates some advance in the hornwort-moss-tracheophyte lineage. Dichotomizing sporophytes are practically unknown except in the tracheophytes, and the apparent lack of stomates and vascular tissue in these simple dichotomously branched axes indicates that they belong to a group of plants totally unknown today but that was probably a part of the stem group of the hornwort-moss-tracheophyte lineage. The small sporophyte axes of *Eohostinella* may have been supported by a **sterome,** a peripheral thick-walled supporting tissue, similar to that described by Edwards et al. (1986). Smooth-walled miospores *(Ambitisporites)* are also known from late Early Silurian (upper Aeronian) strata, as is a second structural type from the Middle Silurian (from the earliest Sheinwoodian). Spores of *Ambitisporites* have well-developed trilete marks, but are otherwise identical to some spores separated from some "permanent" tetrads to which they are probably related. "Permanent" tetrads and dyads, and both alete and trilete monads, are found among living bryophytes. Trilete spores are rare in most mosses (but are known in two families, the Sphagnaceae, where they are common, and in the Encalyptaceae), and also occur in hornworts and liverworts (Boros and Járai-Komlódi, 1975). However, in the fossil record, cryptospores appear before miospores, and they remain dominant in Early Devonian spore assemblages. They are still present in later Devonian assemblages, but their abundance decreases as a proportion of the total sporomorph spectrum. If "permanent" tetrads and dyads represent a primitive feature surviving to the present day, then cryptospores of this type should occur in later geological strata as well.

The expansion in miospore diversity, from the Early Silurian (Llandovery) to Early Devonian (Richardson, 1985; Richardson and Ioannides, 1973; Richardson and Lister, 1969), may reflect diversification of the **rhyniophytoids,** a transitional, though possibly partly artificial group of early evolving higher plants known only as compression fossils and characterized by their small, smooth, dichotomizing axes (but whose anatomy is largely unknown). Rhyniophytoids have both bryophyte and tracheophyte characteristics and probably included ancestors to both.

The primary separation between cryptospores and trilete miospores may also reflect changes in the dispersal mechanisms. In the Late Silurian and earliest Devonian, many miospores and some cryptospores were smaller than 25 μm (the upper size indicating adaptation to wind dispersal in modern bryophytes; Mogensen, 1981), a decrease in size that more or less coincided with the development of erect sporophytic axes. However, although several sporangia in the Late Silurian (Pridolian) contain trilete spores less than 25 μm in diameter, there are others with larger **isospores** (a term meaning that all the spores in a sporangium are similar in size), including cryptospores with thin, frequently collapsed proximal faces (Fanning et al., 1988, 1991). This diversity of spores known from sporangia of sporophytes having similar morphology possibly reflects evolution of the sporophyte, in a single extinct plant group, beyond the level currently found in bryophytes. Alternatively, the apparently similar sporophytes may mask the early stages of diversification of several plant groups. In higher strata of the Pridolian (Late Silurian), *Ambitisporites* has been found within the sporangia in the rhyniophytoid *Cooksonia pertoni*. *Cooksonia* is a genus that includes branched axes with sporangia at their tips (that is, **terminal sporangia**). Possibly the appearance of *Ambitisporites* in the Early Silurian (late in the Aeronian) represents the first appearance of *Cooksonia*-type rhyniophytoid plants.

Gray (1985) considered that the dominant occurrence of trilete miospores could be interpreted as reflecting plants of a vascular level of organization, but from the available fossil evidence, these Silurian plants (that is, those from late Homerian to late Ludlow strata) may have been in the moss-tracheophyte lineage and, thus, in the pre-"true lignin" phase. Silurian rhyniophytoid sporangia lack a central column of tissue, the collumella (present in hornworts and mosses), but the spores in these rhyniophytoids all matured at the same time (that is, spore maturation was *simultaneous* as it is in liverworts and mosses, but a situation unlike that in hornworts in which maturation of the spores occurs over a period of time). Therefore, it seems that Silurian rhyniophytoids have some characters in common with all bryophyte groups, but were nevertheless distinct. **Sterile axes** *(Eohostinella)*, plant axes that lack sporangia, are first found in rocks of the Lower Silurian (possibly belonging to the Telychian, the stage immediately younger than the Aeronian). As we noted, these sporophytes apparently lacked vascular tissue, but like some mosses they may have had a sterome, a jacket of thickened cells in the outer cortex. They were probably epiphytic, growing upon plants of the gametophyte generation of the plant life cycle, and although some axes were dichotomous, other may have resembled unbranched sporophytes of the type seen in modern liverworts, mosses, and hornworts, and in the younger fossil plant *Sporogonites*. Isolated unbranched fossil axes either may be fragments of dichotomizing axes, or may truly reflect their original nature. There is no evidence in *Eohostinella* for vascular tissue, but because these plants appear to have been subaerial sporophytes, it seems likely that they belonged in the stem group of the moss-tracheophyte lineage (that is, that they were both pre-moss and pre-vascular; Mishler and Churchill 1985, Figure 5). Both *Eohostimella* and Silurian *Cooksonia* may have had vascular tissue akin to that in mosses, but there is no unequivocal evidence for this (Edwards et al., 1986).

A second major innovation included in this evolutionary phase, represented by associations of internally banded and smooth tubes, possibly together with a resistant cuticle, is totally unlike any plant living today but possibly reflects the occurrence of lichenlike organisms or, at least, of organisms having a lichenlike strategy. Modern lichens, consisting of associations of algae and fungi, are pioneers of the most inhospitable substrates. The fossil tubes may have formed a structural fabric serving both protective and water-conducting roles for algal filaments (possibly represented by some of the preserved filaments). Taylor (1982) discusses the nature of these fossil associations, and Burgess and Edwards (1988) give beautifully illustrated descriptions of Lower Devonian examples. As Lang has pointed out, there is no doubt that these unusual structures are the remains of land plants. They are found, for example, both in alluvial deposits in the Welsh Borderlands (in the "external facies," that is, those deposits on the periphery of the Old Red Sandstone Continent), and in those of the Caledonian Basin of Scotland (in the "internal, Lower Old Red Sandstone facies," that is, strata from the continental interior).

Evolutionary Phase 3: The *Cooksonia*-Sculptured Sporomorph Phase of the Middle and Late Silurian (late Homerian to early Ludfordian)

The first record of a **fertile (sporangium-bearing) axis**, a dichotomously branched rhyniophytoid, is from late Wenlock strata of Ireland (Edwards and Feehan, 1980) and belongs to the genus *Cooksonia*. At about this same time, cryptospore monads (derived from dyads) and miospores begin to develop proximal, and slightly later distal, radial sculpture patterns. Tracheids and stomates, and spores isolated from individual sporangia, are unknown from rocks of this age. During this phase, spores evolve rapidly in terms of sculpture, and for several structural spore types the distal exine exhibits a well-defined evolutionary series (first laevigate, then verrucate and apiculate; Richardson and Burgess, 1988), possibly as an adaptive response that enabled germination to take place in favorable moisture conditions as Mogensen (1981) has speculated for Holocene bryophytic spores. The simple dichotomously branched sporophytes may have remained epiphytic, and therefore dependent on the gametophyte, but the branching pattern indicates a more advanced character state than that typifying bryophytes (although, in 1935, Bower reported that branching occurs in certain moss sporophytes, and also in some liverworts, which he regarded as aberrant, but which may represent a primitive character, that like xylem and phloem, was subsequently largely lost in modern bryophytes). In any case, the conducting tissue of *Cooksonia* may have been "bryophyte grade," and support for the approximately 0.5-mm-diameter axes of these plants was probably provided by sterome-type cortical tissue.

Evolutionary Phase 4: The Stomata—Rare Vascular Tissue Phase of the Late Silurian (late Ludfordian–early Pridoli)

During this time interval, there was rapid diversification of **homosporous** rhyniophytoids (plants producing only a single size of spore), especially reflected in the morphologies of the sporangia and spores. However, nothing is known about the basal parts of these rhyniophytoids or the nature of their gametophytes. The fossils consist of fragmentary axes, sometimes branched and sometimes occurring as unbranched fragments, of sporangia and of spores. The earliest known stomates occur on sterile axes known from the Late Silurian (Pridolian) of the Welsh Borderland (Jeram et al., 1990). Sporophytes of mosses and hornworts have stomata, but not those of liverworts. In mosses, stomata are commonly concentrated around the base of the sporangial capsule, but not so in these Silurian examples where they are on axes (Edwards, 1989, personal communication). "Sculptured" vascular tissue (Edwards and Davies, 1976), that is, xylem with annular rings or with other "ornamentation," is only rarely seen in Late Silurian compression fossils, although many of the plants of this age may have had vascular tissue of the type familiar in bryophytes, tissue that lacked sculpture on xylem cells and that did not contain "true lignin" (Mishler and Churchill, 1984). A faintly visible differentiated axial area, the zone in which vascular tissue would be expected to occur, is seen in some compression fossils. But because tracheids have not been isolated from these fossil plants it seems possible that although rhyniophytoids may have had a vascular system (both xylem and phloem), this tissue may have lacked "true lignin" (although it is doubtful whether a fossilized compression of a modern moss sporophyte with an unlignified vascular strand would show such differentiation). In petrified material from the Early Devonian Rhynie Chert, two species occur that are of particular interest in this regard: *Aglaophyton major* and *Rhynia gwynne-vaughani*. The former of these had a "mosslike" vascular system, whereas the latter, *R. gwynne-vaughani*, had sculptured xylem and possibly "true lignin." Thus the Late Silurian flora may have included plants representing both types of vascular systems, with those having the "mosslike" organization predominating. As has been pointed out (Edwards and Richardson, 1974), in the absence of plants preserved by petrifaction it is impossible to distinguish between these two morphological conditions.

Remy and Remy (1980) have recorded radially symmetrical gametophytes in the Rhynie flora, and these authors and Schweitzer (1980) have suggested that another fossil plant (*Sciadophyton*), and similar Lower Devonian compression fossils from Canada and Germany, are predominantly gametophytic. Questions therefore remain regarding the status of the sporophyte generation in some of these Early Devonian, and earlier, Silurian, plants.

Sporangia and In Situ Spores. In comparison with sporangia and spores known from earlier sediments, more fossils are known from the Late Silurian, and sporangial shape becomes increasingly varied. Spores of this interval show considerable structural and sculptural variation, suggesting greater plant diversity than is apparent from studies only of sporangial morphology (Fanning et al., 1991). *Cooksonia pertoni* and a new, as yet undescribed plant genus, both first known from this interval, have simple dichotomously branched axes that, in fertile specimens, terminate in flattened, more or less disk-shaped sporangia. However, whereas the *Cooksonia* sporangia are smooth, the sporangia in the new undescribed genus are lumpy and "pustulose." The sculpture patterns on the spores of the two plants are also markedly different. *Cooksonia cambriensis,* a second species of *Cooksonia* known from rocks of this age, has more globular terminal sporangia, but its spores are similar in structure to those of *C. pertoni*. In the latest Silurian (Pridolian), otherwise apparently identical specimens of *Cooksonia pertoni* have structurally similar spores with differing exine sculptural patterns, spores that have been given different scientific names (*Ambitisporites* and *Synorisporites verrucatus;* Fanning et al., 1988). Dispersed spores (that is, spores not occurring within sporangia) of the genus *Ambitisporites* have a long pre-latest Silurian fossil record (being first recorded in the Lower Silurian, upper Aeronian); and, as we noted, they may be derived from "permanent" tetrad-producing ancestors. *S. verrucatus,* in contrast, is known only from the early part of the latest Silurian (the early Pridolian), but similar spores are found somewhat earlier (in the late Homerian), at more or less the same time that *Cooksonia* appears, and also in later pre-latest Silurian (pre-Pridolian) rocks.

Two other discoidal sporangia, similar to the sporangia of *C. pertoni* but not connected to plant axes, have also been studied. One of these contains spores that have a different type of sculpture pattern but that are otherwise similar to trilete *C. pertoni* spores; the other sporangium contains apiculate cryptospores. Thus, both cryptospores and trilete miospores occur in apparently identical sporangia! Either the liverwort-tracheophyte lineage was one group of plants that produced both cryptospores and trilete spores, or the sporangia of several different plant groups were remarkably similar. Whether the two discoidal sporangia were appended to unbranched or branched axes is at present unknown. If these fossil plants included moss, as well as tracheophyte ancestors, then some preservational potential was evidently lost in moss sporangia. Or, alternatively, perhaps these millimetrically small structures have simply been overlooked in younger sediments.

Dispersed Spores. Studies of latest Silurian (early Pridolian) assemblages of dispersed spores have shown them to be dominated by cryptospores (Fanning et al., 1991), but the most abundant cryptospores differ in their sculpture patterns from those of the most abundant dispersed miospores. Many of these miospores are similar to spores occurring in the sporangia either of *Cooksonia pertoni* or of one of the two types of unattached discoidal sporangia just discussed. These studies of dispersed spores only indirectly, and incompletely, reflect the nature of the total vegetation, because the samples are from marginal marine sediments (Richardson and Rasul, 1990) and the spores have been subjected to transport. Nevertheless, they seem to indicate that *Cooksonia pertoni* was an especially important member of the local Late Silurian flora. The apparent lack of Late Silurian sporophytes having cryptospores within their sporangia contrasts markedly with their abundance in dispersed sporomorph assemblages and may reflect the uncharacteristic appearance of those sporophytes which, like the modern hornwort *Anthoceros,* had sporangia that are scarcely distinguishable from their axes.

Miospores. Numerous innovations are found among trilete spores during the Late Silurian (late Ludfordian to Pridoli) interval. Spores with distinctive new sculpture patterns appear, some of which show local variations in abundance that can be correlated with the boundary between the land and the sea. Although other spore types initially increase in abundance away from the shore, as has been shown for the distribution of some modern mangrove pollen (Muller, 1959), part of this increase may be due to the durability and ease of recognition of this particular spore type after prolonged transport (Richardson and Rasul, 1990). In the Late Silurian (Pridolian), trilete spores with a distinctive, distally thick-walled, patinate structure and apiculate sculpture first appear, and it is interesting that of all the sporangia found with spores, none is known to contain patinate spores of this type. Presumably, they are from a distinct group of plants, possibly one having sporangia that are not usually preserved as fossils or having sporangia that because of their small size, or other characteristics, have been overlooked. Alternatively, the plants producing these distinctive spores may not have been transported into the Upper Silurian sediments examined. Numerous other innovations also occurred. In fact, during this interval, the miospore flora evidently underwent a major period of rapid diversification.

Cryptospores. In dispersed assemblages, "permanent" dyads with smooth exines are common, and "permanent" tetrads also occur, as do various other types of cryptospores. Two types of cryptospores have been found within sporangia, one within an elongate fusiform sporangium and the second in an unattached discoidal sporangium of the *Cooksonia*-type. Additionally, several types of alete monads have been reported in rocks of this age.

Evolutionary Phase 5: The Sporophytic Elaboration Phase Beginning in the Early Devonian (Gedinnian–Emsian)

Sporophytic evolution within the tracheophytes involved increases in plant size, branching, and in number of sporangia, as well as changes in tracheid morphology. A large number of small sporangium-bearing axes have been found in Early Devonian deposits, and a wide variety of spores has been extracted from them. These spores provide the potential for differentiating among a whole series of **rhyniophytes** and rhyniophytoids, many of which have morphologically similar sporangia.

In the Anglo-Welsh Basin, assemblages of dispersed spores are known from the Early Devonian (Gedinnian). Included within these assemblages are a large number of variants, referred to the dispersed spore genera *Streelispora* and *Aneurospora*. The distal portion of these spores is covered by an ornament of cones, but they have the same equatorial structure as *Ambitisporites* and *Synorisporites,* spore genera isolated from Late Silurian (Pridolian) specimens of *Cooksonia pertoni.* Early Devonian (Gedinnian) specimens of this rhynio-phytoid species contain spores of the *Streelispora-Aneurospora* complex, thus perpetuating an evolutionary lineage extending back to the Early Silurian (Aeronian). The *Streelispora-Aneurospora* spore complex is diverse and the range of spore variation changes through time, but these spores are rare or absent after the earliest Devonian (latest Gedinnian) and have not been reported from the later part of the Early Devonian (Siegenian) of South Wales. *Cooksonia* fossils that range up to the top of the Lower Devonian (Emsian; Chaloner and Sheerin, 1979) were not closely related to the *Cooksonia pertoni* lineage. According to Edwards and Fanning (Table 1, 1985), no species of *Cooksonia* are known from post-earliest Devonian strata (post-Gedinnian rocks) of Western Europe.

Rhyniophyte and Zosterophyllophyte Lines. In the Early Devonian, two evolutionary lines are apparent, a *Rhynia*-type and a *Zosterophyllum*-type (Banks, 1968). Initially, both lines appear to include rather similar spores, but the spores have different sculptural patterns and apparently develop along different structural pathways. The series proposed are morphological only, and of necessity are tentative, especially because the parallel trends seen in the sculpture of Silurian miospores and cryptospores may be environmentally controlled. Climate may be a factor in determining the morphological "fashions" seen in the Silurian and Devonian spore successions (e.g., proximally tripapillate spores in the Pridolian and Gedinnian; retusoid spores in the Lower Devonian; proximal radial muri in the Lower Devonian; and anchor spines in the Middle and Upper Devonian). Nevertheless, although evolutionary parallelism has probably occurred, there are indications that some spore types may be indicative of particular related groups of plants. Unfortunately, only a few of these genera of dispersed spores have been related with certainty to their parent plants.

Early Devonian rhyniophytes have naked axes that are at first equally dichotomously branched, but in later appearing members of the lineage show an increasing tendency to unequal branching ("overtopping") and an increase in plant size. They have a simple, centrally located cylindrical column of vascular tissue (centrarch xylem, like that, for example, in *Rhynia gwynne-vaughani*), and homosporous terminal sporangia; probably this group was the evolutionary forerunner of ferns and progymnosperms. In contrast to the rhyniophytes, **zosterophyllophyte** axes are both **monopodially branched** (with a single, major axis from which side branches emerge) and dichotomously branched, and both naked or with small leaflike protrusions termed **enations.** They have an "exarch protostele" (phloem has not been seen), and lateral sporangia (with spore sacs attached to the sides of the axes) borne spirally or in rows; probably this lineage gave rise to the lycopods (the present-day club mosses).

Zosterophyllophyte-Lycophyte Line. In Britain, branch systems showing overtopping (unequal dichotomous branching) are known from the Early Devonian (Gedinnian). At this time there was a tendency to develop **lateral sporangia** (Edwards and Fanning, 1985). The first formed xylem elements in these plants (the protoxylem) are located toward the periphery of the elliptical vascular column; thus later-formed elements occur toward the interior of the column, and the xylem is said to mature centripetally. In contrast, in the rhyniophyte line, the first-formed xylem elements are located at the center of the cylindrical vascular column, and xylem maturation is centrifugal (but is known from only one or two examples). Spores known from the sporangia of plants in the zosterophyllophyte-lycophyte line are all trilete and relatively distinctive.

Banks (1968, 1975) derives the **lycophytes** (club moss-type plants) from the zosterophyllophytes based on the concept that the small single-veined leaves **(microphylls)** of lycophytes are evolutionary derivatives of the small bumplike protrusions (enations) occurring on the axes of some zosterophyllophytes (Bower's "Enation theory" of the origin of microphylls). *Asteroxylon,* a well-studied lycophyte-like plant from the Early Devonian Rhynie Chert, has microphyll-like structures but the vascular tissue (the single midvein) stops at the base of the "microphyll." No detailed description has been given of the spores of *Asteroxylon. Leclercqia,* a second well-studied Early and Middle Devonian plant in this lineage (known from the Emsian-lower Givetian), is probably the earliest known fully developed lycophyte. The spores of *Leclercqia* are distinct and easily identifiable; the margin of the contact with the three other spores of the tetrad (the spore curvaturae) is marked by a line of distinct sculptural elements which progressively become more spinose through geological time (Richardson et al., in preparation). The spores walls are characteristically layered (that is, they are **laminate,** consisting of two or three layers distinguishable by light microscopy), and the spores have a thin inner body which is detached in some specimens. The distal sculpture consists of sinuous wall-like elements (muri) surmounted by spinose elements, and is highly variable both within spore "populations" and through late Early to Middle Devonian time (from the middle Emsian to the early Givetian). Dispersed spore "populations" link two species of *Leclercqia,* one in the late part of the Early Devonian (Emsian) and the other later, in the Middle Devonian (earliest Givetian). All of these *Leclercqia* spores can be assigned to the dispersed spore species *Acinosporites lindlarensis.* Another species of *Acinosporites* is known from rocks of about this age and probably belongs to a related type of herbaceous lycopod. Other somewhat similarly sculptured spores (for example, those resembling some species of the genus *Samarisporites*), are also known from deposits of comparable age. The morphological similarities among these various spores, known from Middle and Late Devonian strata (those of the Eifelian, Givetian, and Frasnian), suggest that they may eventually all prove to be from related lycopods.

Rhyniophyte-Trimerophyte-Progymnosperm Line. The **progymnosperms** comprised a large vigorous plant group that appeared during the late Early to Middle Devonian (near the Emsian-Eifelian boundary) and attained their acme in the Late Devonian to Early Carboniferous. They were **free-sporing** (that is, when the spores became mature, they were released from the sporangia into the environment, rather than being retained within "cones"); and they were shrubs or trees, having fronds like ferns but having secondary xylem (wood) like that of gymnosperms.

TABLE 4.1

Summary of Spore Types Described from Sporangia of Rhyniophytoids-Rhyniophytes and Trimerophytes.

Plant Taxa	Spore Taxa
Rhyniophytoids-Rhyniophytes	
Cooksonia pertoni	*Ambitisporites* Fanning et al., 1988
	Synorisporites verrucatus Fanning et al., 1988
	Streelispora newportensis Fanning et al., 1988
	Aneurospora sp. Fanning et al., 1988
Cooksonia cambrensis	?*Ambitisporites/Aneurospora* Fanning et al., in prep.
Caia	*Retusotriletes* cf. *coronadus* Fanning et al., 1990
"Genus nov. P"	*Retusotriletes* sp. Fanning et al., in prep.
Aglaophyton major	*Ambitisporites* (cf. *Perotrilites microbaculatus*) Allen, 1980
Salopella allenii	?Retusoid, ?sculptured outer layer/tapetum (Edwards and Richardson, 1974)
Northia aphylla	*Apiculiretusispora* Bhutta, 1969
Trimerophytes	
Psilophyton charientos	*Apiculiretusispora brandtii* Gensel, 1979
Psilophyton crenulatum	*Apiculiretusispora* cf. *plicata* Doran, 1980
Psilophyton dawsonii	*Retusotriletes* cf. *triangulatus* Banks et al., 1975
	Apiculiretusispora Streel, 1967
Psilophyton forbesii	*Apiculiretusispora brandtii* Gensel, 1979
Psilophyton princeps	*Retusotriletes* sp. Hueber, 1968
	Calamospora atava Hueber, 1968
	Apiculiretusispora Gensel and White, 1983
Pertica dalhousii	*Apiculiretusispora brandtii* Doran et al., 1976
	Apiculiretusispora plicata Doran et al., 1976
Pertica varia	*Apiculiretusispora brandtii* Granoff et al., 1976
	Apiculiretusispora plicata Granoff et al., 1976
Trimerophyton robustius	*Calamospora atava* McGregor, 1973
	Calamospora pannucea McGregor, 1973
	Apiculiretusispora sp. Allen, 1980
	Retusotriletes Allen, 1980

Progymnosperms are thought to have arisen from the **Trimerophytales** (*Psilophyton* and its allies) which, in turn, is a group that may have been derived from the rhyniophytoid-rhyniophyte plexus (Banks, 1968, 1975; Chaloner and Sheerin, 1979). As summarized in Table 4.1, spores have been described from a number of rhyniophytoids-rhyniophytes and trimerophytes, and various of their morphological features may indicate similar evolutionary pathways to those proposed on the basis of plant morphology. **Retusoid miospores** (having a characteristic contact feature on the proximal face) are less common in the Late Silurian (Pridolian) than in later Devonian strata, and although such spores (referred to several species of the genus *Apiculiretusispora*) occur in Late Silurian-age rocks, so far none has been found within sporangia. Two new rhyniophytoid genera (*Caia;* Fanning et al., 1990; and "Gen. nov. P"), known from Upper Silurian (lower Pridolian) rocks of the Welsh Borderland, have similar retusoid spores. Because of their relatively early occurrence and their characteristic morphology, these two genera can be used to illustrate possible pathways in spore evolution (Figure 4.2).

FIGURE 4.2

Proposed evolutionary relationships of various early evolving spore-producing plants (*Cooksonia, Psilophyton, Tetraxylopteris, Rellimia, Archaeopteris, Zosterophyllum Leclercqia*) as reflected by fossil spores: *Tetrahedraletes* (cryptospore) with envelope; miospores, also found in sporangia, with externally laminate exines (the psuedosaccate spore, *Rhabdosporites,* and *Contagisporites*); and internally laminate miospores (*Geminospora, Acinosporites,* and *Ancyropora*). Diagrams illustrate various aspects of wall structure and surficial ornamentation. Two evolutionary lines are illustrated: the Rhyniophyte-Trimerophyte-Progymnosperm line (at left); and the Zosterophyllophyte-Lycopod line (at right). Arrows adjacent to sketches of transverse sections (T.S.) of walls point toward the external surface; "C" denotes cavity.

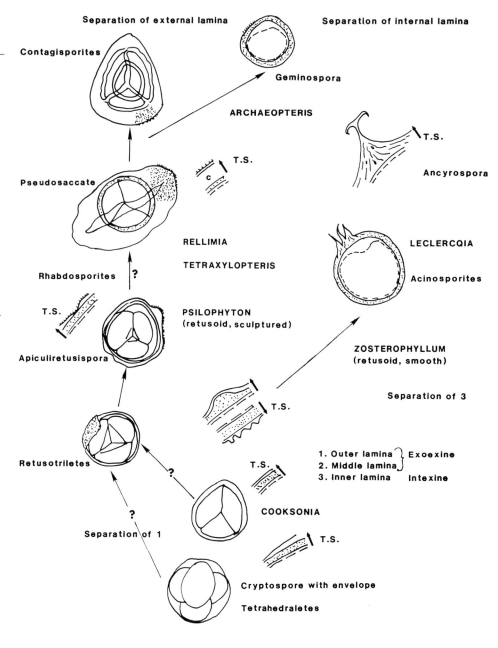

In the earliest Devonian (strata of the lower Gedinnian) of the Anglo-Welsh Basin, a particular type of retusoid spore (*Retusotriletes* cf. *triangulatus*) has a darkened (possibly thickened) apical region. Rare specimens of these spores show a thin, fragmentary outer layer, partially adhering to the spore wall. As shown in Figure 4.2, a morphological series can be developed beginning with spores of this type and leading to **pseudosaccate spores** (those with saclike envelopes only attached proximally to an inner body). The morphological intermediaries are the relatively commonly occurring large spores of the genus *Apiculiretusispora* of the middle to upper Lower Devonian (Siegenian and Emsian strata) that have a darkened apical area and a loose, or partially lost, exterior wall layer (the exoexine) that is covered by numerous, small bumplike protrusions (coni or grana). All the spores of trimerophyte genera so far described (for example, those of *Psilophyton=Dawsonites;* Banks et al., 1975; Gensel and Andrews, 1984; Richardson, 1969) show these features.

The Lower Devonian spores we discussed, in which in many specimens there is a tendency for the outer sculptured layer to separate from the smooth inner layer of the exine and to slough off, are referred to here as **externally laminate.** This organization is morphologically similar to the finely sculptured pseudosaccate spores typical of the progymnosperms in which exoexine is separated from the intexine (the interior wall layer) as a constant feature. Consequently, and although it must be emphasized that Devonian spore structure is not yet adequately known, a hypothetical series of spore morphologies from rhyniophytoids to rhyniophytes, trimerophytes, and progymnosperms can be tentatively proposed. The principal features of this sequence are summarized in Figure 4.2 (viz., from laevigate retusoid spores; to laevigate retusoid spores with a rarely preserved outer apiculate layer; to retusoid spores with a variably separated apiculate exoexine; to pseudosaccate spores with a sculptured exoexine and a smooth, thick-walled inner body). Early in the Middle Devonian (near the beginning of the Eifelian), large spores *(Rhabdosporites langii)* with a sculptured pseudosaccus appear. Identical spores have been found in two morphologically similar progymnosperm taxa, *Tetraxylopteris* and *Rellimia* (Bonamo and Banks, 1966, 1967; Laclercq and Bonamo, 1971).

Evolutionary Trends in the Progymnosperms. In Middle Devonian strata (those of the highest Eifelian or lower Givetian), finely sculptured, internally laminate spores occur (that is, spores with a thick, more or less rigid outer exoexine and a thinner, variably separated intexine) referred to the genus *Geminospora*. The abundance of such spores later in the Devonian (during the Givetian and Frasnian) was such that their occurrence has been termed the "*Geminospora* flood." One of the most common and widespread of such spore species, *Geminospora lemurata*, occurs more or less contemporaneously with the first appearance both of *Aneurospora* (a genus including miospores similar to *Geminospora,* but having banded curvaturae; Streel, 1964) and of *Biharisporites* megaspores (so termed because of their relatively large size). All of these spores have a basically similar wall structure which consists of at least two distinct layers, a thick, sculptured exoexine, and a thinner and variably separated intexine (that is, all are internally laminate). Fine sculptured varieties of these three genera resemble *Rhabdosporites* in being at least two-layered, but differ from this genus by having a more rigid exoexine. Some specimens occurring in Middle Devonian strata (near the Eifelian/Givetian boundary) are difficult to place taxonomically, having been named *Geminospora* by some authors and *Rhabdosporites* by others. The known temporal distributions of fossil plants and spores referred to Devonian progymnosperms are summarized in Table 4.2.

TABLE 4.2	Stratigraphic Distributions of Progymnosperm Fossil Plants* and Spores.

Fossil Plant* and Spore Taxa	Known Stratigraphic Distribution
Rhabdosporites langii	Middle and possibly Upper Devonian (basal Eifelian to upper Givetian, possibly Frasnian)
*Rellimia	Middle Devonian (Eifelian and lower to middle Givetian)
*Tetraxylopteris	Middle and Upper Devonian (upper Eifelian to lower Frasnian)
Aneurospora greggsii	Middle Devonian (lower Givetian to Strunian)
*Aneurophyton	Middle and Upper Devonian (upper Eifelian to lower Famennian)
Geminospora lemurata	Middle and Upper Devonian (possibly uppermost Eifelian, lower Givetian to Frasnian)
Bioharisporites parviornatus	Middle Devonian (possibly uppermost Eifelian, lower Givetian)
Contagiosporites optivus	Middle and Upper Devonian (upper Givetian and Frasnian)
*Archaeopteris	Upper Devonian (Frasnian and Famennian)

*Ranges for plant genera after Chaloner and Sheerin, 1979.

SUMMARY AND CONCLUSIONS

The main points to be gained from this brief discussion of the origin and early evolution of land plants can be summarized as follows:

1. Higher land plants (embryophytes) are composed of two major groups, the bryophytes (represented among living plants by the liverworts, hornworts, and mosses) and the tracheophytes (which have "advanced" conductive tissue composed of xylem and phloem and are represented in the modern flora by, for example, ferns and seed plants).

2. The earliest such land plants had a two-phase life cycle involving the alternation of a gamete-producing phase (the gametophyte generation, producing sperm and egg commonly in separate plants, which ultimately combine), and a spore-producing phase (the sporophyte generation, characterized by its production of haploid, meiotically produced spores which germinate to produce a new gametophytic plant, thus completing the life cycle).

3. The earliest land plants, probably bryophyte-like in overall organization, were derived from freshwater algal ancestors (possibly from charophycean algae); the rise of plant life on land thus required numerous innovations to permit habitation of the harsh subaerial environment.

4. One way to assess the nature of such innovations is to consider the strategies that permit living bryophytes to survive, and thrive, in subaerial environments. Such strategies, possibly similar to those exhibited by members of the earliest higher land plant flora, include short life cycles, drought resistance, control of water loss by anatomical modifications, and photoprotective adaptations.

5. A second way to assess the nature of such innovations is to consider those structural characters that permit tracheophytes (in which the sporophyte generation is dominant) to thrive in subaerial environments. These include the occurrence of spores resistant to

desiccation (a feature present in both bryophytes and tracheophytes); of a waxy cuticle, to prevent water loss from exposed plant axes; of stomata, to permit gas exchange through the desiccation-resistant cuticle; and of vascular tissue, for water conduction (xylem) and nutrient conduction (phloem) as well as to provide physical support (via lignified xylem) to erect axes.

6. Thus investigations of the origin of land plants involve determination of the time(s) of origin of these various adaptations and of the plant group(s) in which they first appeared.

7. Two principal approaches to this set of problems, and to investigation of the evolutionary relationships among early higher land plants, can be used: The cladistic approach, a method that to date has focused chiefly on studies of living plants; and the palynological-paleobotanical approach involving studies of the fossil record.

8. Application of the cladistic method has provided excellent analyses of the relationships among currently living bryophytes, spore-producing plants of the type thought to have been among the earliest members of the land flora. However, because such studies have largely ignored the available fossil evidence, they have provided no insight into the timing of the various events leading to the origin of land plants, nor have they provided direct data regarding the nature of the earliest land flora.

9. In contrast, the palynological-paleobotanical approach has provided much direct evidence regarding the earliest stages in land plant evolution, but because the available fossil record is incompletely known and notoriously incomplete, this approach can provide only a part of the total answer.

10. Clearly, therefore, a synthesis of the two approaches—of data derived both from neobotanically based cladistic analysis *and* from palynological-paleobotanical investigation of the available fossil record—seems necessary to investigate these questions effectively.

11. Cladistic analyses have suggested the following model for the development of the earliest higher land plants:

    a. The archetype of all land plants: A "thalloid" (flat, platelike) gametophyte, with an attached sporophyte consisting only of a sporangium.
    b. The archetype for hornworts, mosses, and tracheophytes: A thalloid gametophyte, with a stomata-bearing sporophyte.
    c. The archetype of mosses and tracheophytes: A radially symmetrical gametophyte, with a vascular strand- (xylem- and phloem-) containing sporophyte having a single elevated sporangium.
    d. The evolution of tracheophytes: In sporophytes, increases in plant size and branching, development of increasingly complex tracheids, and production of leaves.

12. Assessment of the known palynological-paleobotanical fossil record reveals the following five "evolutionary phases" in the early evolution of the land flora:

    Phase 1—the cryptospore-cuticle phase (Ordovician to Early Silurian), characterized by the earliest occurrence of cryptospores ("permanent" spore tetrads and dyads, and separate monads), of sheets of cuticle, and of smooth tubes.

    Phase 2—the sporophyte-trilete isospore phase (Early to Middle Silurian), characterized by the earliest occurrence of dichotomizing (bifurcating) plant axes (e.g., *Eohostinella*), of banded tubes, and of trilete isospores.

    Phase 3—the *Cooksonia*-sculptured sporomorph phase (Middle and Late Silurian), characterized by the earliest occurrence of fertile (sporoangium-bearing) rhyniophytoids *(Cooksonia)* and of spores having sculptured, "ornamented" spore walls.

    Phase 4—the stomata-rare vascular tissue phase (Late Silurian), characterized by the earliest occurrence of plant axes having stomates and (probably lignified) vascular tissue.

Phase 5—the sporophytic elaboration phase (beginning in the Early Devonian), characterized by increases in plant size, branching, and in number of sporangia, by development of increasingly complex tracheids, and by production of leaves.

13. This latter evolutionary phase (phase 5) included such developments as the first appearance of rhyniophytes and zosterophyllophytes, the evolutionary derivation of lycophytes from zosterophyllophytes (in Laurussia), and the evolution of progymnosperms (via trimerophytes) from rhyniophytes.

14. Comparison of the cladist-derived scenario with evidence available in the fossil record reveals the following similarities and differences:
    a. Although the fossil evidence is sketchy, the land plant archetype suggested by cladistic analysis seems consistent with the fossil data represented by evolutionary phase 1.
    b. The archetype for the hornwort, moss, and tracheophyte lineage suggested by cladistic analysis is at least largely consistent with the fossil evidence of phase 2 and, in part, phase 3 (but, as we noted, the relevant portion of the fossil record is as yet incompletely documented).
    c. The archetype for the moss and tracheophyte lineage suggested by cladistic analysis seems to be encompassed by evolutionary phase 4 (and would be particularly so, if the apparent vascular strand of plants of this stage were in fact "unlignified").
    d. The final stage of the cladistic scenario, that of elaboration of sporophytes, is well evidenced by evolutionary phase 5.

15. Because plants exhibit **mosaic evolution,** with different plant parts evolving at different rates (stems faster than roots and leaves; reproductive spores faster than stems and, at times, sporangial shape), some aspects of plants are more useful for studies of evolutionary relationships than others. Plant spores are especially useful, because of their fast, possibly adaptive evolutionary rates and their abundance in the early land plant record.

16. A spore-based evolutionary progression is thus outlined here (Figure 4.2) that ties together the early evolving lineages, from primitive rhyniophytoids *(Cooksonia)* to advanced Middle and Late Devonian progymnosperms.

17. It remains to be determined, of course, whether this spore-based evolutionary succession will stand the test of time; what *is* important is that this scenario is testable, one that will either be confirmed, or refuted, as additional evidence is assembled in years to come.

■

REFERENCES CITED

Allen, J. R. L. 1974. Studies in fluviatile sedimentation: Implications of pedogenic carbonate units, Lower Old Red Sandstone, Anglo-Welsh outcrop. *Geol. J. 9:* 181–208.

Allen, K. C. 1980. A review of in situ Late Silurian and Devonian spores. *Rev. Palaeobotan. Palynol. 34:* 1–9.

Banks, H. P. 1968. The early history of land plants. In E. T. Drake (Ed.), *Evolution and the Environment* (New Haven: Yale Univ. Press), pp. 73–107.

Banks, H. P. 1975. Reclassification of Psilophyta. *Taxon 24*(4): 401–413.

Banks, H. P., Leclerq, S., and Hueber, F. M. 1975. Anatomy and morphology of *Psilophyton dawsonii* sp. nov. from the late Lower Devonian of Quebec (Gaspé), and Ontario, Canada *Palaeontographica Americana 8*(48): 77–127.

Berkner, L. V., and Marshall, L. C. 1965. On the origin and rise of oxygen concentration in the Earth's atmosphere. *J. Atmos. Sci. 22:* 225–261.

Blackmore, S., and Barnes, S. H. 1987. Embryophyte spore walls: Development and homologies. *Cladistica 3*(2): 185–195.

Bold, H. C., Alexopoulos, C. J., and Delevoryas, T. 1980. *Morphology of Plants and Fungi* (4th ed.) (New York: Harper & Row), 819 pp.

Bonamo, P. M., and Banks, H. P. 1966. *Calamophyton* in the Middle Devonian of New York State. *Amer. J. Bot. 53:* 778–791.

Bonamo, P. M., and Banks, H. P. 1967. *Tetraxylopteris schmidtii:* Its fertile parts and its relationships within the Aneurophytales. *Am. J. Bot. 54:* 755–768.

Boros, Á., and Járai-Komlódi, M. 1975. *An Atlas of Recent European Moss Spores* (Budapest: Akademiai), 466 pp.

Bower, F. O. 1908. *The Origin of a Land Flora* (London: Macmillan), 727 pp.

Bower, F. O. 1935. *Primitive Land Plants* (London: Macmillan), 658 pp.

Burgess, N., and Edwards, D. 1988. A new Palaeozoic plant closely allied to *Prototaxites* Dawson. *Bot. J. Linn. Soc. 97:* 189–203.

Chaloner, W. G., and Sheerin, A. 1979. Devonian macrofloras. In M. R. House, C. T. Scrutton, and M. A. Bassett (Eds.), *The Devonian System, Special Papers in Palaeontology 23* (London: Palaeontol. Assn.), pp. 145–161.

Crum, H. 1972. The geographical origins of North America's eastern deciduous forest. *J. Hattori Bot. lab. 35:* 269–298.

Edwards, D., Bassett, M. G., and Rogerson, E. C. W. 1979. The earliest vascular land plants continuing the search for proof. *Lethaia 12:* 313–324.

Edwards, D., and Davies, E. C. W. 1976. Oldest recorded in situ tracheids. *Nature 263:*494–495.

Edwards, D., and Fanning, U. 1985. Evolution and environment in the Late Silurian-Early Devonian: The rise of the pteridophytes. *Phil. Trans. R. Soc. London 309B:* 147–165.

Edwards, D., Fanning, U., and Richardson, J. B. 1986. Stomata and sterome in early land plants. *Nature 323:* 438–440.

Edwards, D., and Feehan, J. 1980. Records of *Cooksonia*-type sporangia from late Wenlock strata in Ireland. *Nature 287:* 41–42.

Edwards, D., and Richardson, J. B. 1974. Lower Devonian (Dittonian) plants from the Welsh Borderland. *Palaeontol. 17*(2): 311–324.

Fanning, U., Edwards, D., and Richardson, J. B. 1990. Further evidence for diversity in Late Silurian land vegetation. *J. Geol. Soc. London 147:* 725–728.

Fanning, U., Richardson, J. B., and Edwards, D. 1988. Cryptic evolution in an early land plant. *Evolutionary Trends in Plants 1:* 13–24.

Fanning, U., Richardson, J. B., and Edwards, D. 1991. A review of *in situ* spores in Silurian land plants. In S. Blackmore and S. H. Barnes (Eds.), *Pollen and Spores, Patterns of Diversification* (London: The Systematics Assn.) (in press).

Gensel, P. G. 1976. *Renalia hueberi,* a new plant from the Lower Devonian of Gaspé. *Rev. Palaeobotan. Palynol. 22:* 19–37.

Gensel, P. G., and Andrews, H. N. 1984. *Plant Life in the Devonian* (New York: Praeger), 380 pp.

Glasgow, L. 1990. The history of the ozone layer. *New Scientist 1744:* 24.

Gray, J. 1985. The microfossil record of early land plants: Advances in understanding early terrestrialization, 1970–1984. *Phil. Trans. R. Soc. London 309B:* 167–195.

Hébant, C. 1977. The conducting tissues of bryophytes. *Bryophytorum Bibliotheca 10* (Vaduz: J. Cramer), 117 pp.

Inoue, H. 1960. Studies in spore germination and the earlier stages of gametophyte development in the Marchantiales. *J. Hattori Bot. Lab. 23:* 148–191.

Jeram, A. J., Selden, P. A., and Edwards, D. 1990. Land animals in the Silurian: Arachnids and Myriapods from Shropshire. *Science 250:* 658–661.

Leclercq, S., and Bonamo, P. M. 1971. A study of the fructification of *Milleria (Protopteridium) thomsonii* Lang from the Middle Devonian of Belgium. *Palaeontographica 136B:* 83–114.

Lemoigne, Y. 1968. Observation d'archegones portés par des axes du type *Rhynia gwynne-vaughani* Kidston et Lang. Existence de gametophytes vascularisés au Devonien. *C. R. Acad. Sci. Paris 266:* 1655–1657.

Markham, K. R., and Porter, L. J. 1978. Chemical constituents of the bryophytes. *Progr. Phytochem. 5:* 181–272.

Mishler, B. D., and Churchill, S. P. 1984. A cladistic approach to the phylogeny of the "Bryophytes." *Brittonia 36*(4): 406–424.

Mishler, B. D., and Churchill, S. P. 1985. Transition to a land flora: Phylogenetic relationships of the green algae and bryophytes. *Cladistics 1*(4): 305–328.

Mogensen, G. S. 1981. The biological significance of morphological characters in bryophytes: The spore. *The Bryologist 84:* 187–207.

Muller, J. 1959. Palynology of Recent Orinoco delta and shelf sediments: Reports of the Orinoco Shelf Expedition, volume 5. *Micropalaeontol. 5:* 1–32.

Pratt, L. M., Phillips, T. L., and Dennison, J. M. 1978. Evidence of non-vascular land plants from the Early Silurian (Llandoverian) of Virginia, U.S.A. *Rev. Palaeobot. Palynol. 25:* 121–149.

Raven, J. A. 1977. The evolution of vascular land plants in relation to supracellular transport processes. *Adv. Bot. Res. 5:* 153–219.

Raven, J. A. 1985. Comparative physiology of plant and arthropod land adaptation. *Phil. Trans. R. Soc. B309:* 273–288.

Remy, W., and Remy, R. 1980. Devonian gametophytes with anatomically preserved gametangia. *Science 208:* 295–296.

Richardson, J. B. 1969. Devonian spores. In R. H. Tschudy and R. A. Scott (Eds.), *Aspects of Palynology* (New York: Wiley Interscience), pp. 193–222.

Richardson, J. B. 1985. Lower Palaeozoic sporomorphs: Their stratigraphical distribution and possible affinities. *Phil Trans. R. Soc. Lond. B309:* 201–205.

Richardson, J. B. 1988. Late Ordovician and Early Silurian cryptospores and miospores from north-east Libya In A. El-Arnauti, B. Owens, and B. Thusu (Eds.), *Subsurface Palynostratigraphy of Northeast Libya* (Libya: Benghazi), pp. 89–109.

Richardson, J. B., and Burgess, N. 1988. Mid-Palaeozoic sporomorph evolution in the Anglo-Welsh area: Tempo and parallelism. *Abstracts, 7th Palynol. Cong. Brisbane, Australia:* p. 140.

Richardson, J. B., and Ioannides, N. S. 1973. Silurian palynomorphs from the Tanezzuft and Acacus formation, Tripolitania, North Africa. *Micropaleontol. 19:* 257–307.

Richardson, J. B., and Lister, T. R. 1969. Upper Silurian and Lower Devonian spore assemblages from the Welsh Borderland and South Wales. *Palaeontol. 12:* 201–252.

Richardson, J. B., and Rasul, S. M. 1990. Palynofacies in a Late Silurian regressive sequence in the Welsh Borderland and Wales. *J. Geol. Soc. London 147:* 675–686.

Schofield, W. B. 1985. *Introduction to Bryology* (New York: Macmillan), 431 pp.

Schopf, J. M., Mencher, E., Boucot, A. J., and Andrews, H. N. 1966. Erect plants in the early Silurian of Maine. *Prof. Paper U.S. Geol. Surv. 550D:* D69–D75.

Schweitzer, H. J. 1980. Die Gattungen *Taeniocrada* and *Sciadophyton* Steinmann im des Rheinlandes. *Bonner Palaeobot. Mitteil 5:* 1–38.

Stewart, K. D., and Mattox, K. R. 1975. Comparative cytology, evolution and classification of the green algae with some consideration of the origin of other organisms with chlorophylls *a* and *b*. *Bot. Rev. 41;* 104–135.

Streel, M. 1964. Une association de spores du Givétien inférieur de la Vesdre, à Goé (Belgique). *Ann. Soc. Geol. Belg. 87:* 1–30.

Strother, P. K., and Traverse, A. 1979. Plant microfossils from Llandoverian and Wenlockian rocks of Pennsylvania. *Palynology 3:* 1–22.

Taylor, T. N. 1982. The origin of land plants: A paleobotanical perspective. *Taxon 31* (2): 155–177.

CHAPTER 5

A HISTORY OF VERTEBRATE SUCCESSES

■

John H. Ostrom†

WHAT IS A "VERTEBRATE" ANIMAL?

The precise evolutionary history of most backboned animal lineages is not known in its entirety for the shocking reason that we simply do not have *all* the evidence. We do not know all of the different kinds of organisms that have ever lived. Yes, we do know about the ancient and modern coelacanths, and the "Nessie-like" seagoing plesiosaurs of the Mesozoic, the earth-quaking *Brontosauruses*, and the dawn horse *Eohippus* and successors *Miohippus, Merychippus,* and *Pliohippus* (Simpson, 1951), and others. But these and the thousands of other ancient backboned fossil species now known give only a fleeting glimpse of the true diversity of life that thrived through the millenia in which backboned animals—those we refer to as **vertebrates** because they possess a vertebral column—first lived and experimented with various lifestyles and modes of existence. Indeed, vertebrates succeeded in adapting to virtually all aquatic niches, whether marine or freshwater, essentially all regions of the terrestrial realm, and the aerial world as well. Nevertheless, their known fossil record is in fact minuscule; over the more than 500 Ma that vertebrates have existed and thrived on this dynamic planet, the fossil record as now known probably includes far less than 1% of all of the kinds of vertebrates that have ever existed.

Since the dawn of Cambrian time and the rise of chordate life, previewed by the fossil *Pikaia* of the Burgess Shale (Gould, 1989), the vertebrate experiment orchestrated one amazing scenario after another: A succession of successes, variations on a single structural theme, the **vertebrate Bauplan.** What is this overall structural blueprint? And why has it been so incredibly successful? Indeed, among all animals, only the arthropods have been more successful (perhaps for similar architectural reasons). In the briefest possible description, the noble vertebrate is a "cephalized, sensate, bilaterally symmetrical, motile, coelomate gnathostome having a segmented endoskeleton, a dorsal hollow nerve chord, and a ventral gut." Most likely, it is this organization and suite of features, this Bauplan, that promoted the successes in the vertebrate story, a succession of diverse experiments in

---

†Division of Vertebrate Paleontology, Peabody Museum of Natural History, Yale University, New Haven, Connecticut 06511 USA

animal form, mode, and capability that occurred throughout virtually the entire Phanerozoic Eon.

THE EARLIEST VERTEBRATES

Animals without backbones, evolutionary precursors of vertebrates, were already well established by latest Precambrian time (see Chapter 3). The oldest undoubted fossil evidence of vertebrate life is represented by small fish fossils, having the name *Astraspis,* preserved in the Middle Ordovician Harding Sandstone of Colorado. *Astraspis* specimens most commonly are frustratingly fragmentary, usually just angular chips of calcium phosphate, mineral material similar in chemical composition to the bones of most later vertebrates. Even older similar fragments, but of less certain affinity, are known from Late Cambrian-age rocks of Wyoming and Oklahoma.

We might consider the occurrence of *Astraspis* (or of the other similar forms) as representing the first major event in the history of vertebrates, but that would overlook their progenitors, their evolutionary sources. Among known *non*vertebrates is an innocuous array of animals classified as **chordate** (because, like vertebrates, chordates have a dorsal nerve chord), but which lack a true *segmented* backbone. Instead, these animals feature an internal, flexible, axial rod for support, the **notochord,** a structure considered to be the evolutionary precursor of the segmented backbone of vertebrate animals and a structure that in fact occurs in the early growth stages of many modern vertebrates. The living **cephalochordate** amphioxus *(Branchiostoma),* with its elongated, bilaterally symmetrical body, anterior "head" region, posterior finlike tail, and most of the other attributes we listed earlier, is an ideal anatomical model from which the vertebrate Bauplan could easily have been derived. However, as a modern, living organism, amphioxus is nearly a half billion years too young to be a vertebrate progenitor! Nevertheless, we do know that an amphioxus-like animal actually did exist during Middle Cambrian time, perhaps 530 Ma ago (Gould, 1989). This remarkable fossil, from the famed Burgess Shale of British Columbia, Canada, christened *Pikaia* by its discoverer Charles D. Walcott, is known from only a few dozen specimens (Figure 5.1A); but those elongated silvery impressions reveal segmented anatomy strikingly similar in shape and size to that of the modern amphioxus (Conway Morris and Whittington, 1979). *Pikaia* may well represent a very early stage among chordates that led to the vertebrate design; as such, this fossil would constitute evidence of the *first* major stage, the earliest recorded event, now known in vertebrate history.

The Advent of Fish

Astraspis and a number of other small fish of varying shapes termed **ostracoderms** were evidently rather abundant in the Late Ordovician and Silurian seas (Denison, 1967). Although they are known only from fragments or impressions, three characteristics appear common to all. First, the fossil remnants consist entirely of external or superficial head/thoracic shields, or of a mosaic of tiny body scales. None shows evidence of any internal skeleton. Presumably, the internal skeleton was cartilaginous (like the amphioxus notochord) and simply not preserved. Second, none of the fossils shows any signs of true paired fins or appendages. And third, none of these earliest vertebrates (if they can be appropriately so termed) had a lower jaw! The mouth existed only as a simple opening rather than as a more complex, hinged structure. For this reason, these animals have been grouped together in a single category known as the **Agnatha,** the jawless fishes. Although poorly

FIGURE 5.1

Pikaia (A), a probable cephalochordate (protochordate) similar to the modern amphioxus (B), from the Middle Cambrian Burgess Shale of western Canada. Animal length is about 40 mm (photograph courtesy of S. Conway Morris).

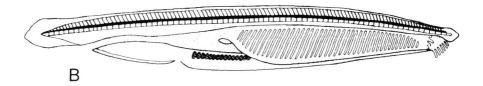

B

known and not at all well understood, these primitive fish nevertheless represent a significant advance beyond the amphioxus-like, cephalochordate design.

Two subsequent major additions to the vertebrate plan were the development of paired appendages or fins, and the evolution of a true lower jaw, hinged against the skull or head armor. These two events happened independently and at different times, but both were advantageous innovations that apparently happened more than once. Some agnathids, such as the **cephalaspids,** bore paired, crude "pectoral fins" that probably aided in stabilizing the body against currents and in maneuvering during swimming. Although the cephalaspid paired appendages are completely unrelated in evolution to the anterior paired limb configuration of later evolving fish or other vertebrates, their adaptive value for stabilizing and controlling movements is evident. Similarly, a hinged lower jaw has distinct advantages over the jawless condition for the all-important tasks of food gathering and processing. The fossil evidence indicates that lower jaws developed, perhaps independently, in several different groups of early fish, among these the small spiny-finned **acanthodians** of the Silurian; the larger and more ferocious-looking **placoderms** of the Devonian; the remarkably successful (if less than popular) **sharks,** that also date from the Devonian; and the true **bony fish,** the ray-finned and lobed-finned fish, of later Paleozoic, Mesozoic, and Cenozoic times. These two novelties, paired fins and a lower jaw, acquired early in vertebrate history, are important attributes in virtually all subsequent evolving vertebrate lineages; clearly, jaws and limbs became essential components of vertebrate organization.

One other important early alteration in the vertebrate Bauplan must be noted. The earliest fish had external or superficial (dermal) head plates and shields, and body armor in the form of small, thick scales, but showed little or no sign of an internal skeleton—that is, they had no distinct vertebral column, ribs, pectoral or pelvic limb skeleton, and had little in the way of skull ossifications beneath their superficial dermal head shields. The evolutionary precursors of such internal supports must have been cartilaginous.

Development of the Internal Mineralized Skeleton

At first glance, it seems puzzling that the cartilaginous supports of early evolving fish were evidently supplanted by ossified internal skeletal architecture in later evolving fish, a development reflected in the bony skeletons of their terrestrial descendants, the amphibians, reptiles, birds, and mammals. Clearly, in the weightless world of the sea, an internal cartilaginous equivalent (or modification) of the primitive notochord provided a sufficient framework against which segmented body muscles contracted to effect the necessary undulatory swimming action. Thus there seems to have been no obvious need or advantage for the internal skeleton of fish to have become ossified. But there must have been good reasons for superficial ossifications to develop in the form of head and thoracic shields, and body scales. Perhaps the principal advantage was protection, a shielding against some aspect of the external aquatic environment. But protection from what? From parasites? Algae? Deleterious radiation? From predatory eurypterids? For whatever reason, the aquatic fauna of the world underwent a conspicuous change from mid-Paleozoic to Tertiary times, a major increase in the diversity of bony fish (**actinopterygians**) characterized by increasingly ossified internal skeletons and by a general reduction in external ossification, and a marked decrease in the number of kinds of fish that featured heavy scale coverings. This pattern seems to hold for the fish community in general, cutting across phyletic lines (except for cartilaginous sharks and their relatives, and for certain other, unusual adaptive specialties). A simplistic explanation could be that an internal ossified skeleton, in contrast to a covering of thickened, external body scales, may have enhanced body flexibility and promoted more efficient muscular action (by facilitating muscle contraction on alternate sides against the vertebral column) required for the undulatory swimming motion.

THE RISE OF ANIMAL LIFE ON LAND

Ossification of the internal skeletal framework was an essential precursor for the next major event in the history of vertebrate life, invasion of the land surface. Without a sturdy internal skeleton, the first vertebrate invaders would have been hard pressed to withstand the force of gravity. The gravitational obstacle involved in the transition from the weightless world of the seas, lakes, and rivers to the world of dry land is only one of the several barriers that had to be overcome, but an obvious barrier that was perhaps the most critical. Part of the solution to this problem is well evidenced in the fossil record, documented by the increased ossification of the internal skeletal framework of certain bony fish, in particular those termed lobe-finned fish, or **coelacanths.** The probable evolutionary progeny of the lobe-fins of the Devonian and Carboniferous were the first true amphibians (Figure 5.2), namely, the ichthyostegids, the anthracosaurs, and the temnospondyls (groups also known collectively as **labyrinthodonts**). Subsequently, by the early or mid-Mesozoic, the more familiar kinds of amphibians, the frogs, toads, and salamanders, made their appearance. But the labyrinthodonts were evidently the earliest amphibians, the first vertebrate explorers of the world's land areas and, therefore, the first to confront the problem of gravity on the land surface and the additional difficulties of breathing, hearing, and seeing while out of water. They also faced the life-threatening dilemma of desiccation, a loss of body fluids, as they left the security and relative comfort of their home waters and began exploring the shore banks. These earliest pioneers should be recognized as key players in the cast of characters at this crucial stage in the history of vertebrate life, the first of several stages in vertebrate "colonization" of the land surface, an environment previously occupied by plants and some invertebrates, but not by higher animals. To the newly evolved, colonizing vertebrates, this was a new empire to be explored and exploited.

Origin of the Amniote Egg

Successful invasion of the land was possible only because of the newly acquired internal skeletal system that resisted gravity, supported body weight, and provided (via paired

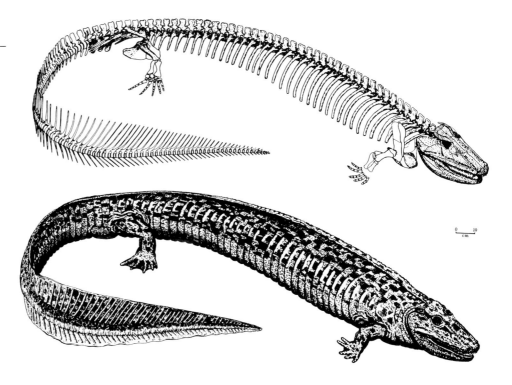

FIGURE 5.2

Eogyrinus, an Early Pennsylvanian anthrocosaur, a typical labyrinthodont amphibian, as restored in skeletal and living appearance. Labyrinthodonts were the original vertebrate explorers of the land. Total animal length was about 4 m (from Panchen, 1972, with permission).

limbs, in essence, converted fins) the necessary leverage and propulsive strokes to push the early labyrinthodonts across the beaches to the inland frontiers. The fossil record does not provide much insight as to why these animals left the security of the water, how they coped with the resulting respiratory, visual, or auditory problems, or how they dealt with the hazardous dangers of fluid loss.

Nevertheless, at least some of these pioneers managed to deal successfully with these obstacles. But they still faced yet one more critical dilemma, namely, in this new, hostile environment, *how could they reproduce?* In a marine or freshwater setting, the aquatic medium is a natural "helpmate," providing means for transit of the fragile gametes (sex cells) to necessary but improbable union. Removed from the shelter of the aquatic environment, the need to reproduce in the dry environs on land presented a major challenge. Fortunately, at least one successful strategy evolved, development of the **amniote egg,** a hard-shelled, self-contained system capable of both protecting, and nourishing, the developing embryo (Figure 5.3). This evolutionary "invention," occurring in some unknown subgroup of the labyrinthodont lineage, solved the critical problem that until then had effectively stymied vertebrate success on land.

FIGURE 5.3

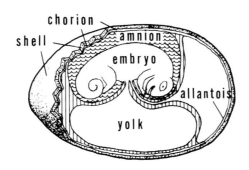

Diagrammatic cross section of an amniote egg with major components labeled. Notice that the developing embryo occupies a completely self-contained environment that provides food (yolk), respiration ability (allantoic sac), anti-desiccation protection (amniotic sac), and physical shelter (chorion and shell).

In addition to the advents of nerve cords, fins, jaws, limbs, and the like, the origin of the amniote egg is one more major event—among the most important of all such events—in the history of vertebrate life for the very simple reason that it finally and completely opened up the terrestrial frontier, making it possible for vertebrates to have permanent reproductive independence from their aquatic origins. From the hindsight of the present day, development of this freedom must be considered *the* most important event in vertebrate history after adoption of the overall vertebrate Bauplan via the cephalochordate blueprint previewed by *Pikaia* (Figure 5.1). Indeed, the origin of the amniote egg in a struggling labyrinthodont (Figure 5.2) released the extraordinary evolutionary potential of the vertebrate genome.

Were it not for several critical aspects and potentials of the vertebrate Bauplan, it is entirely unlikely that there could ever have been such exquisite creatures as swans, hummingbirds, Thompson's gazelles, black-footed ferrets, performing porpoises, or racing thoroughbreds, to say nothing about the thinking, comprehending *Homo sapiens*. Without the remarkable evolutionary invention of the amniote egg, neither you would be here to read this, nor I would be here to write it. Most vertebrates are **oviparous** (meaning that their eggs mature outside of the maternal body), and their eggs are therefore subjected to the irregularities and hazards of potentially hostile environments. On the other hand, **ovoviviparity** (in which the developing egg is retained within the female) is not uncommon in "lower" vertebrates, occurring in some fish, amphibians, lizards, and snakes. A specialized version of ovoviviparity in which the fetus is nourished within the mother **(viviparity)**, as in placental mammals, represents the pinnacle of vertebrate reproductive success. All of these reproductive strategies are based on evolutionary derivatives of the original amniote egg.

The Earliest Amniote Vertebrates

From the "first" amniote egg hatched a peculiar labyrinthodont-like "pup" that we can imagine must have exceeded all of its parents' expectations. It was, in fact, by traditional definition, a **reptile;** that is, it belonged in the same biologic category as lizards and snakes, turtles and crocodiles. These living examples are mentioned here only to identify for you the few important kinds of reptiles that have survived to the present. However, also included in the reptile category are extinct "sea monsters" (the plesiosaurs, mosasaurs, and ichthyosaurs), the dinosaurs, flying reptiles (the pterosaurs), and so-called mammal-like reptiles (technically, the Synapsida) that fossil evidence indicates are the direct ancestors of our own class Mammalia. Thus the direct ancestors of mammals were of reptilian stock; they reproduced by means of the amniote egg; and they were scaly, rather than hairy. The live-birth placental strategy of the mammalian grade had not yet been perfected. Theoretically, it seems just a small step from oviparity (known to occur in a few varieties of living reptiles) to the placental mode of reproduction characteristic of mammals.

Although the best known of all fossil reptiles, the dinosaurs, are established to have been egg-layers (Brown and Schlaikjer, 1940; Granger, 1936) rather than live-bearing creatures, they were enormously successful. It is essential to emphasize here that the success of the dinosaurs was only *one more example* of the many successful vertebrate experiments that can be attributed to the vertebrate Bauplan and the novel amniote egg that made untapped terrestrial real estate available to new occupants. The fortunate beneficiaries of the amniote strategy and vertebrate scheme include all reptiles, both living and extinct, plus all reptilian descendants, both avian and mammalian. Clearly, the origin of the amniote egg was an extraordinarily important event—the penultimate major event in the history of vertebrate life. To demonstrate its importance, let us review some of its evolutionary consequences, a series of developments that led up to the origin of mammals, the ultimate major event in vertebrate history.

Evolution of Early (Anapsid) Amniotes

The "first" amniotes are not clearly recognizable in the fossil record for the obvious reason that remnants of such animals and their eggs have not been found fossilized together. Even if they were, we could never be certain that the very first-evolving example had been discovered. Moreover, probability dictates that the first amniotes were very rare, occurring only locally, and were thus unlikely to have been preserved in the rock record. The early evolutionary radiation of amniotes may have taken place toward the beginning of Carboniferous time or even before (judging from the report of an unnamed "amniote" from Lower Carboniferous strata of Scotland), and quite likely the amniotes produced by this first phase in the evolution of the group were small vertebrates that ambled about in the dense lycopod forests of that time. Early Pennsylvanian (Late Carboniferous) reptilelike forms are known from such fossilized forests, preserved in hollow tree stumps that perhaps served as their terrestrial habitats, or at least as temporary shelters. Collectively, these forms are called **captorhinids;** they were lizardlike in general body proportions and appearance, with short necks and heads, long tails, and short- to medium-length sprawling legs (Carroll, 1969). These primitive earliest known amniotes are distinct from all living reptiles, except turtles, in the primitive construction of their skulls which lack cranial openings (**fenestrations**) in the temporal region. Captorhinids thus exhibit the so-called **anapsid** condition. The absence of temporal openings is a reflection of the primitive state of the jaw apparatus and its musculature. Advanced jaw mechanics and feeding specialties had not yet arisen, and powerful jaw muscles seem not to have been required. Presumably, the captorhinids preyed on insects abundant in the Late Carboniferous forests, and therefore had no need for powerful jaws. Best known of the captorhinids are small lizardlike forms of the genera *Hylonomus* (Figure 5.4) and *Paleothyris,* both of Early Pennsylvanian age. These and related primitive amniotes seem to have proliferated in later Pennsylvanian and Permian time. They serve as a model, or **archetype** of the primitive amniote condition with which we compare all subsequent "reptilian" vertebrates, of which there were a great many varieties.

■

THE EVOLUTIONARY PATH TO MAMMALS

Synapsids—Primitive "Mammal-like Reptiles"

One such lineage, derived early from the captorhinids, demonstrates a very early trend among terrestrial vertebrates, a trend toward specialization in the business of food gathering and processing. Evidence for this trend comes from fossil tetrapod remains contemporane-

FIGURE 5.4

Skeletal reconstruction of *Hylonomus,* one of the earliest known amniotes ("reptiles"), from the Early Pennsylvanian of Nova Scotia, Canada. The animal length is about 25 cm (from Carroll and Baird, 1972, with permission).

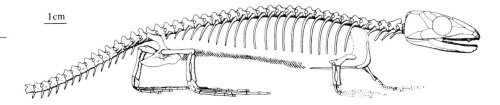

1 cm

ous with *Hylonomus* (in fact, occurring with *Hylonomus* in the same rock strata of Nova Scotia), members of the fossil genus *Protoclepsydrops.* Unlike captorhinid skulls, the temporal region of *Protoclepsydrops* skulls are perforated by distinct fenestral openings indicating the presence of enlarged jaw muscles. Other remains from mid-Pennsylvanian strata, in particular *Archaeothyris,* show that this early cranial fenestration was no fluke. These latter fossils constitute the earliest evidence of relatively advanced jaw muscles, another evolutionary novelty, and they therefore have been referred to a distinct new subclass of reptiles, the **synapsids.** As such, they are the earliest known "mammal-like reptiles," the progenitors of our own class, Mammalia.

It may seem strange that two distinctly different major kinds of amniotes—anapsids and synapsids—could have both appeared so early in the historical record, but perhaps it is not. The amniote condition opened the broad vistas of the terrestrial environment to whatever applicants were at hand with the right kinds of necessary equipment. Whereas the anapsid captorhinids were evidently able to persist mostly on inherited primitive equipment, as relatively small animals preying on even smaller insects, the hypothetical (as yet unknown) "proto-synapsids" appear to have had a degree of genetic plasticity (not foresight) that enabled them to develop a broader mouth gape, enlarged jaw muscles, and specialized dentitions, adaptations that in turn permitted them to prey on larger animals. Appearing early in the Pennsylvanian, these postulated primitive synapsids evidently radiated and diversified, filling a variety of predaceous niches. Ultimately, by the end of the Carboniferous, synapsids were abundant and widespread. Among the most distinctive were the **pelycosaurs** of the Carboniferous and Permian, typified by such well-known genera as *Ophiacodon* and *Dimetrodon.* It should surprise no one that these early synapsid tetrapods expanded and spread over the landscape, to occur in almost all parts of the globe, because their talents were focused on *feeding;* they were well adapted for snaring prey of all sizes and processing their kill efficiently with slicing, stabbing, and tearing teeth powered by large, strong jaw muscles.

The evolutionary radiation and wide dispersal of pelycosaurs might be termed the first true vertebrate invasion of the lands of the earth, but by Late Permian time that invasion had waned, perhaps due to the increasing climatic harshness of the terrestrial world, or to competition with a rising new breed of even more advanced mammal-like reptiles, the **Therapsida.** Thus, after their very successful excursion onto the landscape, the pelycosaurs declined and became extinct, leaving behind the results of a new synapsid experiment, the therapsids, perhaps originally derived from a primitive pelycosaur such as *Haptodus.*

Like the pelycosaurs before them, therapsids also features the synapsid cranial condition, although with greatly expanded jaw muscles and highly specialized, well-differentiated teeth. The expanded jaw muscles of the therapsids were more effectively arranged than in pelycosaurs, providing an even wider gape and more powerful bite action. Therapsids rapidly diversified, giving rise to both small and large varieties (some much larger than any previous land animal), to both carnivores and herbivores, and perhaps to some omnivores. The earliest therapsids appeared during the Late Permian, about the time that pelycosaurs were on the decline.

For a number of reasons, the radiation of therapsids during the Late Permian to Triassic interval properly can be regarded as another major event in the history of vertebrate life. First and foremost, these pelycosaurian descendants were more advanced than the pelycosaurs in many ways: Anatomically, posturally, dentally, physiologically, and perhaps reproductively. Second, the therapsid radiation was rapid and widespread, the vast majority of therapsid fossils occurring in the "southern continents" (i.e., the modern continental fragments of the giant southern landmass, Gondwanaland, of late Paleozoic to early Mesozoic time). Third, the therapsid radiation led to very great diversity, resulting in many more new varieties than had the previous pelycosaurian radiation. Moreover, the vast numbers of therapsid specimens now known, such as those found in southern Africa, for example, suggest that therapsids were much more numerous than their predecessors as well.

Finally, among members of the therapsid radiation was included the immediate ancestor of *all* mammals. Thus, from a human (if chauvinistic) point of view, we must regard the rise of the therapsids as one more major event in the history of vertebrate life!

Therapsids—Advanced "Mammal-like Reptiles"

Therapsids commonly are subdivided into two major categories: The abundant and very strange-looking herbivorous **anomodonts,** and the obviously mammal-like, carnivorous **theriodonts.** The anomodont radiation began in mid-Permian time with a primitive lineage, the **Eotitanosuchia,** members of which resembled certain (sphenacodont) pelycosaurs but lacked their elongated neural spines and had a somewhat more upright, slightly less sprawling, posture. Shortly after the appearance of the eotitanosuchians, the more advanced **dinocephalians** arose, a group including such awkward-looking animals as *Moschops* and *Titanosuchus* of the Late Permian. Unlike the posture of pelycosaurs and eotitanosuchians, in which all four limbs are similarly sprawled, the dinocephalian posture was "half and half," with long sprawling forelimbs and shorter, more erect hind limbs. Presumably, this strange posture provided more effective movement than the sprawling style, although it certainly looks ungainly.

Perhaps the most successful of the numerous anomodont groups was the **dicynodont** lineage, a very diverse group of strange "big-tusked" herbivores that arose in the Late Permian and persisted throughout most of the Triassic. The success of the dicynodonts apparently derived from their unusual feeding adaptation. Ranging from squirrel size to the size of a rhino, nearly all were "two-tuskers," most having only a single pair of upper tusks (canines) remaining in the jaws, all other teeth having been lost. In place of the missing teeth, the jaws were equipped with sharp-edged horny beaks, continuously growing, self-sharpening cutting edges useful for cropping all kinds of plant material. The head of a dicynodont was short-faced and boxlike, with enlarged temporal fenestrae for the enlarged, rearranged jaw muscles that powered cropping. Some dicynodonts were at least semi-aquatic, like *Lystrosaurus* (a genus essentially ubiquitous on southern landmasses); others apparently were burrowers (for example *Oudenodon*); and other huge forms were browsers (e.g., *Kannemeyeria* and *Stahlecheria*).

The other major category of therapsids, the mammal-like theriodonts, also underwent an evolutionary radiation beginning in the mid-Permian. Three major groups or lineages radiated: the **gorgonopsians,** the **therocephalians,** and the **cynodonts.** Gorgonopsians were small- to moderate-sized carnivores that featured enlarged upper canine tusks, strong incisors, and greatly reduced cheek teeth. Their adductor jaw muscles apparently were large, as indicated by their enlarged temporal fenestrae, to facilitate the wide-jaw-gaped stabbing action of their large canines. Their body posture appears to have been semierect— near-sprawling forelimbs combined with an intermediate half-sprawl, half-upright position in the hind limbs. These gorgonopsians were contemporaneous with (and sometimes in competition with) members of the second radiating lineage, the more advanced therocephalians. Although generally similar to gorgonopsians in having large upper canines and reduced cheek teeth, therocephalians were more diverse in size and in specialized carnivorous adaptations. Their jaw muscles were expanded over those of the gorgonopsians, and they exhibited other novel mammal-like cranial innovations such as a secondary palate that probably permitted simultaneous breathing and food processing.

The radiation of the third group of theriodonts, the cynodonts, began in the Late Permian, peaked during the Triassic, and subsequently dwindled, with the group becoming extinct by the mid-Jurassic. Of all therapsid radiations, the cynodont radiation is, without question, the most important, not only because of its greater duration and possibly greater diversity, but also because of its evolutionary progeny—the class Mammalia.

Cynodonts, the Precursors of Mammals

Because it led ultimately to *Homo sapiens,* identification of the cynodont radiation as another major event in the history of vertebrate life is understandable and therefore a topic that deserves more than cursory review. Fossil remains indicate that the therocephalians and gorgonopsians that dominated the Late Permian landscape were overcome and supplanted by the more advanced cynodonts during the Triassic. Why did this occur? In retrospect, it seems evident that the answer is because these highly advanced therapsids achieved the largest suite of mammalianlike conditions and features. In this regard, it should be noted that (1) most of these mammal-like features are directly or indirectly related to the activity of feeding—food procurement, processing, and ingestion (and perhaps digestion as well); (2) these innovations were made possible only because of the earlier development of the synapsid condition; and (3) the innovations involved significant changes in cynodonts in posture, limb position, mode of movement, and, implicitly, in activity levels and metabolic rates.

Let us consider these numerous cynodont mammal-like features, starting with the teeth. All cynodonts (see Figure 5.5) have well-differentiated teeth: Incisors, canines, and multicusped cheek teeth (premolars and molars) of intricate design for oral preparation of varied food substances in a variety of complex ways. The complex cusp patterns on opposing counterpart teeth, occluded in increasingly precise fashion, enable the crushing, shearing, piercing, and shredding of different food material at different tooth positions. During the cynodont radiation, the cynodont lower jaw became increasingly dominated by enlargement of a single component, the tooth-bearing dentary bone. Accordingly, the main muscles of jaw closure (adduction) shifted in position to become attached more and more on the dentary bone, increasing the bite force and the precision of tooth occlusion. Similarly, the jaw muscles in cynodonts became greatly enlarged as the temporal openings expanded and the lower temporal arch (the zygoma) bowed outward. The original, simple, solitary jaw adductor muscle (the externus) of the early theriodonts ultimately separated into two muscles, evolutionary precursors of the superficial (masseteric) and the deeper (temporalis) jaw muscles of mammals. One of these superficial muscles, the masseteric, was attached to the zygomatic arch (beneath the synapsid fenestra), and the other, the temporalis, originated on the side of the braincase (within the fenestra). Bony support of a secondary palate made possible simultaneous breathing and chewing/swallowing, with full separation of the air passage from the food tract. The postdentary bones became reduced to tiny splinters, precursors of the tiny, bony ear ossicles of the mammalian auditory system.

Other changes occurred in limbs, posture, and related features. The hind and forelimbs shifted to become more and more beneath the hip and shoulder sockets, permitting the more nearly vertical (parasagittal) limb excursions in fore and aft arcs required for more efficient and rapid locomotion. As a result, the process of walking or running became less and less of

---

FIGURE 5.5

Thrinaxodon, an advanced Early Triassic cynodont therapsid from the Triassic of South Africa, as reconstructed by F. A. Jenkins (1971). Animal length about 60 cm.

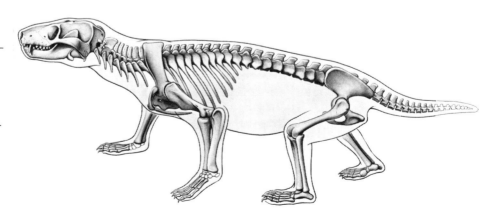

an undulatory, side-to-side waddling, and more of a dorsoventral arching and stretching of the back, coupled with fore and aft limb movements. In some advanced cynodonts, the backbone was modified to restrict the degree of lateral flexibility (necessary in earlier therapsids for their sinuous side-to-side gait), and to provide the greater vertical flexion of the back necessary to facilitate the vertical stride pattern. And in many cynodonts there was a reduction or complete loss of the posterior ribs, and a clear distinction of thoracic and lumbar regions of the body (an organization suggesting the possible presence in advanced cynodonts of a mammal-like diaphragm and, thus, of a more efficient respiratory system). These numerous changes are well illustrated by comparing the various features in primitive, intermediate, and advanced cyodonts (e.g., *Procynosuchus, Probainognathus,* and *Thrinaxodon;* Figure 5.5). All these developments, in concert, suggest that during the cynodont radiation, these animals progressively became capable of increasing physical activity and, therefore, exhibited increasingly higher metabolic rates.

The therapsids were firmly entrenched in the terrestrial habitats of the world during Late Permian to Early Triassic time. As we noted, many members of the group, but particularly the cynodonts, were becoming increasingly mammal-like. Indeed, there is no doubt whatever that therapsids ultimately gave rise to mammals. Thus, it seems surprising that the fossil record has not recorded a smooth, continuous passage from the therapsid-dominated fauna of the early Mesozoic to the terrestrial mammalian faunas of the Tertiary and today. But that was not the way things happened. Another major event in the history of vertebrates got in the way. That event was the advent of archosaurs, specifically, the dinosaurs.

■
_____

DINOSAURS AND THE MID-MESOZOIC EVOLUTIONARY GAP

Way back in the early days of amniotes, during the Early Pennsylvanian, the first two experiments of "reptileness" occurred—development of the first anapsid, *Hylonomus,* and of the earliest synapsid, *Protoclepsydrops.* But a third experiment was also underway at about that time. In Late Pennsylvanian strata at Garnett, Kansas, we find the first known **diapsid** vertebrate. (In contrast to the single pair of temporal openings in synapsids, diapsids featured two pairs of temporal fenestrae; these, too, are correlated with the size, shape, and organization of the head muscles that powered the lower jaws.) The oldest diapsid, *Petrolacosaurus* (Figure 5.6), is known from dozens of specimens from the famous Garnett site in Kansas. *Petrolacosaurus,* like its slightly earlier evolving amniote competitors, was more or less lizardlike, but it was slightly larger (Reisz, 1977). In fact, it may have been the progenitor of all later diapsids, including all **archosaurs,** the major grouping that includes today's crocodiles, alligators, and their relatives and ancestors, plus all dinosaurs, flying reptiles (pterosaurs), and a group of primitive Triassic animals commonly called **Thecodontias.** Other nonarchosaurian diapsids include today's lizards, snakes, the tuatara *(Sphenodon)* of New Zealand, and a few other extinct forms of late Paleozoic and Mesozoic age.

FIGURE 5.6
_____

Reconstruction of *Petrolacosaurus* skeleton, the earliest recognized diapsid reptile, from the Late Pennsylvanian Garnett locality, Kansas. Animal was about 60 cm long (from Reisz, 1981, with permission).

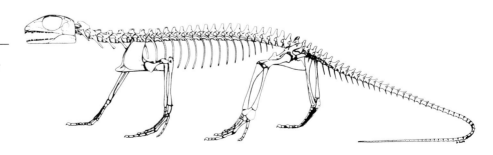

The important fact to be recognized is that the archosaurs—the dinosaurs, in particular—disrupted what otherwise would have been a smooth, continuous, uninterrupted evolutionary progression, a history of higher vertebrates that would have read "from pelycosaurs (Sphenacodontia) to therapsids, from advanced therapsids (Cynodontia) to true placental mammals." Although this progression did, in fact, occur, it was by no means "smooth, continuous, and uninterrupted." Indeed, there was a major temporal gap in the fossil record sequence, an evolutionary hiatus that spanned most of the Mesozoic (almost 150 Ma in duration) during which the landscapes of the world were dominated by dinosaurs. That dinosaurian dominance—not the dinosaurs per se, but their impact on the evolutionary progression—must be registered as yet one additional major event in the history of vertebrate life.

Everyone knows that dinosaurs are commonly classified into two major categories, the **Ornithischia** (those with a birdlike pelvis) and the **Saurischia** (those with a reptilelike pelvis). Most authorities suspect that these two groups had a common ancestor within the order Thecodontia, but that progenitor has not yet surfaced. What follows here is a brief review of the two major dinosaurian categories and some of their most conspicuous features so that we can consider the possible reasons behind the enormous success of the dinosaurs and, thereby, better ponder the "why and wherefore" of the curious Mesozic gap in the history of therossids (i.e., of "mammalness").

"Bird-Hipped" Dinosaurs

The order Ornithischia is perhaps the more primitive of the two major groups. It included a wide variety of dinosaurs, all of which were herbivorous. The definitive feature of all ornithischians is the birdlike construction of the pelvis, in which the pubic bones have two prongs or processes (an anterior blade, and a posterior projection situated parallel to the ischial bones, the latter an arrangement similar to that in modern birds). Hence, this order of archosaurs carries the name Ornithischia, a name meaning "bird-hipped." All ornithischians also featured a unique midline bone, the predentary, that united the lower jaws. The earliest and perhaps the most primitive of the ornithischians are those called **ornithopods** (a name meaning "bird foot," despite the fact that their feet were not particularly birdlike). Nearly all of these ornithopods were **facultative bipeds**—that is, they had the ability to walk on either two feet or four feet—even though most of them probably preferred a quadrupedal gait. Well-known ornithopods were animals such as *Iguanodon, Hypsilophodon,* and the duckbilled dinosaurs, to cite a few. Bipedal dinosaurs of the **Pachycephalosauria,** the "dome heads," are sometimes recognized as distinct from the ornithopods, although the two groups were anatomically very similar. Other nonornithopod ornithischians, such as the well-known plated dinosaurs (**Stegosauria**), armored varieties (**Ankylosauria**), and the various horned dinosaurs (**Ceratopsia**; Figure 5.7), were obligate quadrupeds. The temporal distribution of these various ornithischians is interesting: The ornithopod experiment extended from the mid-Triassic to the latest Cretaceous; stegosaurs were confined to the Early Jurassic to Early Cretaceous interval; ankylosaurs occurred from the Late Jurassic to the latest Cretaceous; and the horned dinosaurs were entirely restricted to the Cretaceous (with nearly all such ceratopsians occurring only during the last half of Cretaceous time). It is important to emphasize the different time ranges characteristic of each group; in general, dinosaur stock changed dramatically through time, with each kind being typical for its time, but *not* for all times.

Judging from their preserved teeth, all ornithischians appear to have been herbivorous. Tooth form varied in detail from one group to another, but in all ornithischians, teeth were leaf-shaped with crenulated margins. In stegosaurs and ankylosaurs, the teeth were quite small and the tooth rows very short, indicating that these dinosaurs were probably sustained on soft plant material. Neither stegosaurian nor ankylosaurian teeth occluded directly; thus

FIGURE 5.7

Skeletal reconstruction (A) and actual skull (B) of *Triceratops*, a common North American ornithischian herbivorous dinosaur of Late Cretaceous time. Animal length was about 9 m. B=#1964 I458, Bayersche Staatssammlung für Palaontologie, Munich, Germany, formerly Yale Peabody Museum # 1834 (from Ostrom and Wellnhofer, 1986).

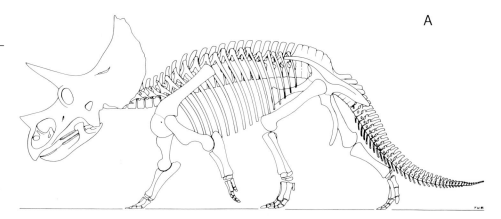

neither of these groups included "good chewers." However, and quite surprisingly, both some ornithopods and all ceratopsians did have the highly specialized dental systems required for chewing! This specialization is unexpected because living reptiles do not chew their food. Mammals are able to chew, but *not* reptiles. Why is this the case? Simply because reptiles, with their relatively lower rates of metabolism, do not require the same amount of nourishment coupled with rapid digestion and assimilation as mammals.

Inasmuch as all dinosaurs are classified as reptiles, the presence of highly specialized chewing dentitions in ceratopsians and in advanced types of ornithopods raises some important questions. (In contrast, it is interesting to note that we are not at all surprised at the complex tooth crowns for chewing that are the hallmarks of advanced mammal-like reptiles, the cynodonts!) In the case of the duckbilled (**hadrosaurian**) ornithopods, the occlusal wear facets form long, inclined grinding/crushing surfaces that are constructed of hundreds of teeth. In the horned (ceratopsian) dinosaurs (Figure 5.7), the occlusal wear facets form long, vertical shearing surfaces, also composed of hundreds of teeth above and below. Clearly, these differences in dentition must reflect a great difference in the food preferences of the hadrosaurian and ceratopsian consumers. But both obviously required large quantities of well-chewed food. In contrast, members of the stegosaurian and ankylosaurian lineages seem to have required far less food, that was evidently unchewed. Why are there these differences in the food requirements of various dinosaurian stocks? Perhaps, as we discuss later, they *might* be related to another, even more basic difference, the possibility that some dinosaurs were "cold-blooded," like modern reptiles, whereas others may have been more nearly "warm-blooded," and thus more "mammal-like" in their metabolism.

"Reptile-Hipped" Dinosaurs

The other major category of dinosaurs, the order Saurischia (those with a reptilelike pelvis), is noted for its inclusion of the largest of all known dinosaur varieties, the giant **sauropods** such as *Apatosaurus* (formerly known as *Brontosaurus*). Also included in the Saurischia are the very different theropod dinosaurs, all of which were flesh eaters, and all of which were obligate bipeds. The huge, massive sauropods were all plant eaters and obligate quadrupeds, a logical posture for such gigantic creatures, most of which weighed tens of tons! But despite their enormous bulk, the dentition of sauropods was not particularly well suited for chewing or preparing food. There is no question that these animals were herbivores, and they certainly required vast quantities of food, but how they processed their food is simply not known. Nor do we know how they even swallowed amounts of food adequate to sustain such huge bodies. Yet we can be certain that they required much more food than does the modern African elephant, the largest now-living land vertebrate, a mammal that comes equipped with huge grinders plus a handy and effective "food picker."

The remaining group of saurischians, the **Theropoda** (which means "beast foot," even though their feet are far more birdlike than mammal-like), are the carnivorous dinosaurs; included here is the "tyrant lizard," *Tyrannosaurus,* on one extreme, and the much smaller chicken-size *Compsognathus,* on the other. These theropods are so different from the sauropods that we cannot help wondering why the two groups are classified together as saurischians. Largely, they are so grouped because both have the saurischian reptilelike pelvic condition, but it is very likely that they should be placed into separate orders. Indeed, such a separation seems logical when we consider their anatomy.

The theropods are especially interesting and distinctive in that they were all **obligate bipeds.** None of the Theropoda was able to assume the quadrupedal stance typical of virtually all nonavian vertebrates. Aside from birds, there are very few vertebrates that are limited to a mandatory bipedal pose. Moreover, bipedality requires a very high degree of neural development, balance, and substantial agility. One of the most remarkable of all theropods, the Early Cretaceous *Deinonychus* (Figure 5.8), illustrates this requirement dramatically (Ostrom, 1969). As is shown in Figure 5.8, *Deinonychus* carried its killing weapons, large sicklelike talons, on its feet—on the feet of an *obligate biped!* There can be no doubt that *Deinonychus* was an extremely active and agile predator. It is difficult to imagine a better equipped and more aggressive assailant!

Like the various ornithischians, the varieties of saurischians also ranged through their own particular time periods. The sauropods were most abundant and widespread in the Jurassic, declining to very few kinds by the Late Cretaceous, while their presumed ancestral stock, the prosauropods, a group of often facultatively bipedal dinosaurs that had dominated the Triassic interval, disappeared altogether. As a group, the theropods thrived from the mid-Triassic to the latest Cretaceous, but no single theropod persisted for more than a fraction of that time.

It must be stressed that throughout the Mesozoic there was a continual turnover of dinosaurian kinds—of *all* kinds: At differing times, the various groups originated, rose to dominance, thrived, and then declined; new groups supplanted the old; previously successful groups ultimately became extinct. As Tennyson wrote in *Idylls of the King:* "The old order changeth, yielding place to new. . . ."

A

Artist's reconstruction (A) and actual skeletal replica mount (B) of *Deinonychus,* currently displayed in the Yale Peabody Museum. Animal length about 3 m (from Ostrom, 1969).

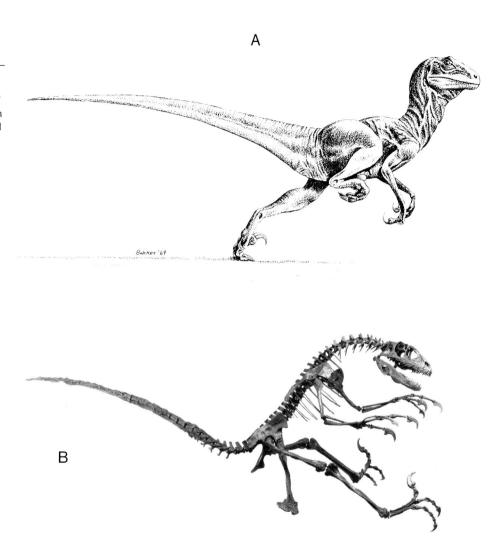

B

The Mid-Mesozoic Historical Gap

Let us now return to that curious Mesozoic gap in the evolutionary progression from therapsids to mammals—to what extent were dinosaurs the cause of this "discontinuous evolution"? Whether or not all types of dinosaurs were particularly closely related to each other (which seems unlikely), as a total group the dinosaurs obviously dominated the lands of the earth for a very long time—from the mid-Triassic to the very end of the Cretaceous—an interval of nearly 150 Ma. Why they were so dominant is not entirely clear, but there may be some clues. It is not likely that there was much, or perhaps even any, significant direct competition for space or food between Triassic to Jurassic dinosaurs and coexisting mammal-like cynodonts. Two facts are evident, however: Both therapsids and archosaurian dinosaurs were advanced anatomically, and both seem to have been very active types of animals. In fact, it has been theorized that both therapsids and dinosaurs may have been "warm-blooded," with high metabolic rates. Certainly, this seems reasonable for the therapsids, because of their probable evolutionary position ancestral to mammals, and the evolution of the group toward increasing "mammal-likeness" with an upright posture and highly complex, differentiated teeth. Both their posture and dentition point to high food requirements and thorough food processing, factors that obviously suggest animals of high metabolic rates and activity. But most dinosaurs *also* had upright posture, and some featured very sophisticated dental equipment. Thus, like therapsids, dinosaurs have been proposed to have been endothermic animals having high metabolic rates, a possibility consistent with their anatomy, their upright posture, their mammal-like bone microstructure, and the ratio of predatory dinosaurs (theropods) to nonpredatory types as preserved in certain Late Cretaceous strata (Thomas and Olson, 1980).

Implicit in the possibility that some dinosaurs may have been "warm-blooded" is the notion that these types may also have been physiologically superior to the early Mesozoic therapsids, a superiority permitting the archosaurs to displace large therapsids and forcing early mammals into small-animal niches. Indeed, it is striking that no mammals larger than a house cat (with virtually all being much smaller) are known until after the close of the Cretaceous. Another point to be made is that dinosaurian locomotion and posture seem to have been superior to those of all contemporaneous therapsids (presumably illustrated, for example, by the obligatory bipedality of *Tyrannosaurus* and all other theropods), perhaps reflecting contrasting or noncompetitive physiologic regimes.

Recently, a number of other interesting and very relevant facts have come to light that should be included in these deliberations. Back in the 1920s, an American Museum of Natural History expedition in Mongolia established that at least some dinosaurs (exemplified by *Protoceratops*) were egg-layers rather than live-bearers. Much more recently, this egg-laying reproductive strategy has been linked with other dinosaur varieties as well (Horner, 1982). Even more importantly, evidence uncovered by Horner clearly indicates that, not surprisingly, some dinosaurs followed a well-defined reproductive procedure, returning to the same breeding and nesting grounds year after year to deposit their clutches. And Horner's discoveries even indicate that the patterns of nested hatchlings seem to show evidence of parental care (Horner and Makela, 1979).

Horner's finds and interpretations are all the more impressive in view of other quite independent evidence, dinosaurian fossil footprints (Figure 5.9) from newly appreciated sites around the world that reveal previously unknown aspects of dinosaur behavior. The vertebrate paleontologist Roland Bird first cast light on such evidence of possible social behavior among dinosaurs in his 1939 letter to *Natural History* and his subsequent article "Did Brontosaurus ever walk on land?" (Bird, 1944). He believed that dinosaur footprint evidence from Texas indicated yes, *"Brontosaurus"* (i.e., *Apatosaurus*) could, indeed, lumber across the land surface. But Roland Bird saw much more—the evidence, he

---

FIGURE 5.9

Footprint evidence of herding behavior in sauropod (A) and ornithopod (B) dinosaurs. (A) From Davenport Ranch, Texas (after Bird, 1944); (B) at Holyoke, Massachusetts (from Ostrom, 1972).

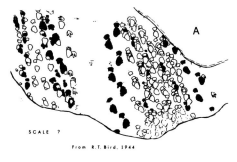

From R.T. Bird, 1944

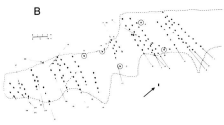

thought, showed herding behavior among sauropods! His conclusion has now been substantiated by evidence from dozens of footprint sites around the world that has been obtained by many investigators for many kinds of dinosaurs (Gillette and Lockley, 1989; Ostrom, 1972). It thus appears that dinosaurs of many kinds were much more sophisticated in their behavior, even to the extent of being more social animals, than they have been given credit for in the past.

Ectothermy, Endothermy, and the Paleophysiology of Dinosaurs

To properly compare the physiology of therapsids on one hand and dinosaurs on the other, in order to assess the cause of the Mesozoic gap in the evolutionary progression we have discussed, it is necessary first to consider some general physiologic definitions. In common parlance, most vertebrates can be described as *cold-blooded,* an imprecise label commonly applied to fish, amphibians, and reptiles. The term is imprecise because it has been applied to a wide variety of animals having very different thermal regimes and requirements, varying from quite low to, in some cases, surprisingly high body temperatures. A more appropriate term for discussing the thermoregulatory state of cold-bloods is the term **ectothermy;** conversely, warm-bloods are better referred to as **endotherms** (Figure 5.10). In brief, ectothermy is that pattern of thermoregulation in which body temperature depends chiefly on behaviorly regulated uptake of heat *from the environment* (as in most amphibians and reptiles, among living land vertebrates). Endothermy, on the other hand, is that pattern of thermoregulation in which body temperature depends on a high *internally* controlled rate of metabolical heat production. In other words, in endotherms, body temperature is generated internally as needed, or as a by-product of exercise and high metabolic processes, quite independent of external environmental temperatures. In contrast, ectotherms rely on external sources of heat (solar radiation, high temperatures) and behave accordingly. Consequently, ectotherms have relatively lower metabolic levels and, therefore, have lower

FIGURE 5.10

Diagram contrasting ectothermic (alligator, snake, lizard) and endothermic (pigeon, cat, rabbit) modes of thermoregulation in vertebrates.

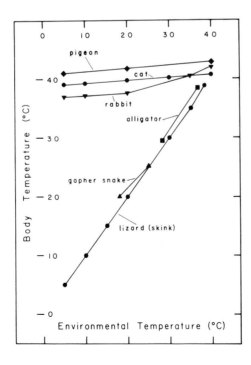

food needs to satisfy those requirements. The bottom line is that the higher rate of metabolism necessary to maintain an endothermic system requires more fuel than an ectothermic system of the same size. For this reason, endothermic mammals or birds require the intake of approximately ten times more food calories per day than do ectothermic lizards or turtles of the same size.

With this background, it is easy to appreciate the economics of ectothermy versus endothermy, and the relative costs of reptilelike versus mammal-like physiology, crucial matters involved in pondering the Mesozoic therapsid-to-mammal evolutionary gap. This is not to suggest that one of these strategies is necessarily "better" than the other. Both strategies exist; both do the job; but they do that job in different ways. Over the short haul a good ectothermic biologic "machine," the southwestern whip lizard *(Cnemidophorus)*, for example, can outrun a perfectly good endotherm, the well-known road runner *(Geococcyx),* but an ectotherm lacks the stamina necessary to stay in front of an endothermic predator for very long. The now well-publicized hypothesis that some (or, by some accounts, all!) dinosaurs were endothermic (Bakker, 1986), and therefore capable of high exercise metabolism, is a very intriguing idea, one that could go a long way toward explaining the notable mid-Mesozoic gap in theropsid history. (It must be noted here that fossil mammal remains are, in fact, known from this curious mid-Mesozoic gap, but that the fossils are not abundant, and that all of these mammals were very small; few exceeded chipmunk or squirrel size.) But just exactly what caused the evidently discontinuous nature of the transition from the dominant therapsid faunas of the early Mesozoic to the later, Tertiary dominance of true mammals, is not apparent in the fossil evidence. The numerous lines of evidence now available (erect posture; bone histology; predator-prey ratios; dental food-processing capabilities in ceratopsians and hadrosaurs; and the probable existence of four-chambered hearts in long-necked dinosaurs like sauropods, ornithopods, and theropods) are all suggestive, but *none* is conclusive. For the time being, the matter must remain unresolved, a question to be answered by future investigations.

The Origin of Birds

Although the issue is unresolved, it is probably too early to dismiss the possible impact that the dinosaur community may have had on therapsid evolutionary history. Indeed, there is yet one more parameter that needs to be considered. We may suspect that thermoregulation (and its associated physiological advantages) was important in Mesozoic organisms, just as it is in modern creatures. That notion has been the root of the presumption that advanced therapsids, those in the evolutionary pathway leading to placental mammals, were endotherms with high metabolic levels approaching those of their modern endothermic descendants. A new line of evidence, however, has changed this picture. In addition to mammals, there is only one other living endotherm group: Birds, members of the **avian** class. Over the past decade, a large body of evidence has been recognized that points to the dinosaurian origin of birds (Ostrom, 1976). Specifically, the skeletal anatomy of *Archaeopteryx,* the oldest well-documented bird (Figure 5.11), is in virtually all respects more dinosaurian than avian; specifically, the bony architecture of *Archaeopteryx* closely resembles that of *theropod* dinosaurs (Hecht et al., 1985). Curiously, as has been noted several times, were it not for the impressions of feathers preserved in several of the specimens of *Archaeopteryx*, these fossils would almost certainly have been identified as reptilian, particularly theropod, rather than as being avian. (In fact, three of the known six specimens of *Archaeopteryx* were originally misidentified, and two of them were actually ascribed to the coelurosaurian theropod *Compsognathus*!) There now seems to be a consensus that birds are the direct descendants of certain dinosaurs of the small theropod clan Coelurosauria. So, in a sense, the evolutionary, phylogenetic scales are balanced: Synapsids led to endothermic mammals, but saurischian archosaurs led to endothermic

FIGURE 5.11

Berlin specimen of *Archaeopteryx*, the oldest undoubted fossil bird, from the Late Jurassic Solnhofen limestone of Bavaria, Germany. Notice the theropod-like anatomy of the skeleton, which is the size of a crow (from Ostrom, 1976).

birds. Today, birds and mammals (Figure 5.12) dominate terrestrial vertebrate communities. These two present-day successes within vertebrate history had their beginnings in Early Pennsylvanian time, one via the synapsid experiment and the other by way of the Late Pennsylvanian diapsid innovation. Because both of these successes involve endothermy, it is interesting to speculate that they may both have been consequences of the development of highly effective (but differing) thermoregulatory/high metabolic regimes.

The drama we have outlined—the origin of amniotes and the rise of reptiles; the origin of mammals and the contemporary rise of dinosaurs; the apparent conflict between the two, with the initial dominance of dinosaurs and their ultimate demise, followed by the radiation of mammals from the earliest amniote stock; and birds rising like the Phoenix from the fallen saurians—all this played to a diverse, packed Mesozoic community including a lush flora of higher plants. At least three major events occurred, the final act being played in three scenes: First, a dramatic shift in the floras of the world about mid-Cretaceous time; second, the Late Cretaceous demise of the dinosaurs; and third, the Cenozoic rise of mammals. The mid-to-late Cretaceous shift from a gymnosperm-dominated terrestrial flora to a flowering plant-(angiosperm-) dominated landscape like that of the present day undoubtedly had a major impact on all of the herbivorous "players," especially on those comprising the ornithischian and sauropod communities. This major floral change may also have provided an initial impetus for the rise of early placental mammals.

Why Did the Dinosaurs Become Extinct?

This Mesozoic drama leaves us, while perhaps not breathless, at least with a sense of wonderment at all of the action that led to the modern world and its occupants. At the same time, we are left with a sense of being unfulfilled, the knowledge that the story is not yet complete, that there is a remarkable amount that yet remains to be learned. The modern assemblage of surviving vertebrates reminds us of at least one remaining mystery, that of dinosaur extinction and the crises confronted by many other life-forms at the end of the Cretaceous. It would be satisfying to know what brought about the massive end-Cretaceous extinction event, a turnover in the biota that changed the terrestrial faunal and floral communities of the world forever.

FIGURE 5.12

Icaronycteris, a 50-Ma-old fossil bat specimen from the early Eocene Green River Shale of Wyoming. The specimen is more than 99% complete and is modern in virtually all respects. Bats represent one of several pinnacles in mammalian history and are perhaps the most remarkable of all living mammals. Specimen is YPM - PU # 18150. Scale bar equals 10 mm.

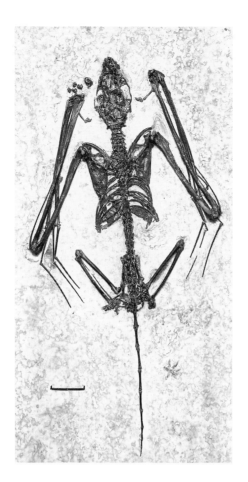

The faunal and floral crises at the end of the Cretaceous, including the notorious demise of the last remaining dinosaurs, have stimulated perhaps more thought than almost any other topic within natural science. What *did* bring about the extinction of the dinosaurs? Let me here state my personal opinion, a point of view that may not be shared by everyone. Personally, I do not think that the dinosaurs crashed all at once at the end of Cretaceous time. There is much too much evidence that over the total span of dinosaur existence, from the mid-Triassic to the latest Cretaceous, there had been one dinosaurian turnover after another. Yet, there is a growing school of believers that latest Cretaceous dinosaurs as well as other members of their contemporary biota, were victims of some major, perhaps extraterrestrially caused, catastrophe. I happen to believe that repeated dinosaurian faunal turnovers were related to **plate tectonics** (the movement of massive portions of the earth's crust across the surface of the globe) and a continuously changing world geography with resulting changes in global climatic patterns. Moreover, although toward the end of the Cretaceous dinosaurs certainly were of declining abundance and diversity, such decreases, just as clearly, were *not* characteristic of *all* terrestrial life.

There is ample evidence (specifically, the famed spike in abundance of the element iridium at the Cretaceous-Tertiary boundary), from many sites around the world, that the earth was impacted by a large extraterrestrial object—an asteroid, a comet, or a meteor swarm—at, as best can be determined, precisely the end of the Cretaceous. Evidence for such an event is compelling; and it is quite conceivable that such infall may have contributed to the massive faunal and floral turnover at the end of the Cretaceous. However, the selectivity of the survivors (dinosaurs died, yet crocodiles, turtles, and higher plants survived) confirms that the episode was not quite that simple; the whole answer is not yet at hand.

Clearly, following this last great extinction, there were enough mammal and bird, and gymnosperm and angiosperm survivors to repopulate the post-Mesozoic world. We must marvel at the ingenuity and durability of life.

■

REFERENCES CITED

Bakker, R. T. 1986. *The Dinosaur Heresies* (New York: Morrow), 481 pp.

Bird, R. T. 1944. Did *Brontosaurus* ever walk on land? *Nat. Hist. 52* (2):60–67.

Brown, B., and Schlaikjer, E. M. 1940. The structure and relationships of *Protoceratops*. *Annals New York Acad. Sci. 40* (3):133–265.

Carroll, R. L. 1969. Origin of reptiles. In C. Gans (Ed.), *Biology of the Reptilia* (London: Academic Press), pp. 1–44.

Carroll, R. L., and Baird, D. 1972. Carboniferous stem reptiles of the family Romeridae. *Mus. Comp. Zool. Bull. 143:* 321–363.

Conway Morris, S., and Whittington, H. B. 1979. The animals of the Burgess Shale. *Scientific American 240* (January): 122–133.

Denison, R. H. 1967. Ordovician vertebrates from western United States. *Fieldiana Geol. 16:* 131–192.

Gillette, D. D., and Lockley, M. G. (Eds.), 1989. *Dinosaur Tracks and Trails* (New York: Cambridge Univ. Press), 454 pp.

Gould, S. J. 1989. *Wonderful Life* (New York: Norton), 347 pp.

Granger, W. 1936. The story of the dinosaur eggs. *Nat. Hist. 38:* 21–25.

Hecht, M. L., Ostrom, J. H., Viohl, G., and Wellnhofer, P. (Eds.). 1985. *The Beginnings of Birds* (Eichstatt, Germany: Freunde des Jura-Museums), 382 pp.

Horner, J. R. 1982. Evidence of colonial nesting and site fidelity among ornithischian dinosaurs. *Nature 297:* 675–676.

Horner, J. R., and Makela, R. 1979. Nest of juveniles provides evidence of family structure among dinosaurs. *Nature 282:* 296–298.

Jenkins, F. A. 1971. The postcranial skeleton of African cynodonts. *Yale Peabody Mus. Bull. 36:* 1–216.

Ostrom, J. H. 1969. Osteology of *Deinonychus antirrhopus*, an unusual theropod from the Lower Cretaceous of Montana. *Yale Peabody Mus. Bull. 30:* 1–165.

Ostrom, J. H. 1972. Were some dinosaurs gregarious? *Palaeo. Palaeo. Palaeo. 11:* 287–301.

Ostrom, J. H. 1976. *Archaeopteryx* and the origin of birds. *Biol. Jour. Linnean Soc. 8:* 91–182.

Ostrom, J. H., and Wellnhofer, P. 1986. The Munich specimen of *Triceratops* with a revision of the genus. *Abhandl. Bayer. Staats. Pal. Hist. Geol. 14:* 111–158.

Panchen, A. C. 1972. The skull and skeleton of *Eogyrinus attheyi* Watson (Amphibia: Labyrinthodontia). *Phil. Trans. R. Soc. London. 163B:* 279–326.

Reisz, R. 1977. *Petrolacosaurus*, the oldest known diapsid reptile. *Science 196:* 1091–1093.

Reisz, R. R. 1981. A diapsid reptile from the Pennsylvanian of Kansas. *Special Pub. 7* (Lawrence, KS: Univ. Kansas Mus. Nat. History), 74 pp.

Simpson, G. G. 1951. *Horses* (New York: Oxford Univ. Press), 247 pp.

Thomas, R. D. K., and Olson, E. C. (Eds.). 1980. *A Cold Look at the Warm-blooded Dinosaurs*. AAAS Symposium 28 (Boulder, CO: Westview Press), 514 pp.

CHAPTER 6

MAJOR EVENTS IN THE HISTORY OF MANKIND

■

Phillip V. Tobias†

■

INTRODUCTION: WHAT IS "MANKIND"?

Without qualification, the term **mankind** is as vague as the term *man*. It means different things to different people. Unless we offer more precise terminology and rules of usage, scientists, philosophers, and laypeople are liable to be at cross-purposes in discussions on mankind. It is therefore necessary at the outset to explain the different uses of the term and to define the sense in which the term is used here.

In one usage, the term *mankind* refers to the members of the zoological family of man, which, following the proposals in 1825 and 1827 of John Edward Gray has been called the **Hominidae.** Species and individuals that are included in the Hominidae are known colloquially as hominids. This is the broadest connotation of mankind, and the one most widely used in discussions on man's origin and evolution.

In a second usage, the term is confined to members of the genus *Homo.* This Latin term, *Homo,* was applied to the varieties of living mankind by Carl von Linné (or Linnaeus) in the tenth edition of his *Systema Naturae,* in 1758. Later researches, in the 19th, and especially the 20th century, have shown that the Hominidae included at least one other genus apart from *Homo.* Hence, if we confine the term *mankind* to those classified in the genus *Homo,* we are using it in a somewhat restricted sense.

Third, the most restrictive mode of use of mankind is its application only to the living species, *Homo sapiens,* so designated by Linnaeus (1758). The term tends to be used in this specific sense by those who are concerned with differences of behavior and mentality between modern humans and the extant apes, particularly with regard to such human characteristics as creativity and ethical and philosophic systems. It is, however, generally not used in this narrow sense by students of human biology, paleoanthropology, and archeology.

†Director, Palaeo-anthropology Research Unit, Department of Anatomy and Human Biology, University of the Witwatersrand, Johannesburg, South Africa

For this chapter, the broadest usage of the term *mankind* is the most appropriate: Accordingly, the following discussion is devoted to a consideration of major events in the history of a group of higher primates, members of the Hominidae.

The members of three groups of higher primates, namely, humans, the great apes (chimpanzee, gorilla, and orangutan), and the lesser apes (gibbons and siamangs), share so many morphological features as to justify the view, now long accepted, that these groups belong to a single major unit in the classification of the mammals and of the primates. To this major unit, the American paleontologist and taxonomist George Gaylord Simpson assigned the rank of a superfamily, to which in 1931 he gave the name **Hominoidea.** So the term hom*inoids* includes gibbons, great apes, and humans, whereas the name hom*inids,* as used here, refers only to humans in the wide usage.

It is important to stress that the classification of living creatures depends basically on their *morphology,* their bodily structure. In the animal kingdom, morphology includes all aspects of bodily structure, both macroscopic and microscopic. It embraces bones and brains, muscles and organs, nerves and blood vessels, as well as tissues, cells, and their minute contents including such structures as chromosomes and mitochondria. Increasingly, in recent decades, other aspects of the **phenotype** (the actual traits of organisms), such as blood types, serum groups, and a variety of identifiable molecules, have been pressed into service as aids to the classification of living forms. Unfortunately, in the study of fossils we are denied the evidence of chromosomes, the microscopical bearers of the hereditary material, of which the members of each living species possess a distinctive complement. Furthermore, except in more recent and some subfossil remains, direct biochemical evidence, for example of blood and serum groups, or of collagen types, is generally unavailable to students of evolution. Hence, we are obliged to use different lines of evidence when we study ancient fossilized remains, compared with those used in the investigation of recent and modern forms of life. An outline of these various research strategies follows.

■
———————

THREE RESEARCH STRATEGIES

Strategy One: The Phase of Hominid Emergence

The major research strategies in the study of hominid origins comprise (1) analysis of the paleoanatomy of fossils; (2) comparison of such fossils with the morphology of related living forms, taking into account available chromosomal (cytogenetic) data; and (3) molecular biological study of living hominids *(Homo sapiens)* and of nonhominid primates.

Strategy Two: The Further Evolving of Established Hominids

In the untangling of evolutionary patterns within the Hominidae from about 4.0 to 1.8 **Myr B.P.,** * the study of the paleoanatomy of hominid fossils is the major part of the research strategy. This study of anatomy is supplemented by three essential sets of data, those on dating, on paleoecology, and from about 2.5 Myr onward, on archeology. Cultural

———————

*Myr B.P. = millions of years before the present.

evidence plays an ever-increasing role in the unraveling of human evolution. The paleoanatomy of the fossils is thus supplemented and enriched by the parallel testimony of material culture and ethological insights, and inferences that may be drawn from them regarding such aspects as group size, distribution, food habits, technical and rational intelligence, language ability, and the world of ideas of the evolving hominids.

Strategy Three: The Phase of Modern Human Evolution

In this phase, which is essentially the evolution of mankind in the Pleistocene epoch, we depend on (1) anatomical, functional, and developmental analysis of fossil and modern human populations; (2) the molecular biological study of various populations of modern human beings; and (3) the concepts and data of archeology, social anthropology, ethnology, linguistics, and ethology.

MAJOR EVENTS IN THE EMERGENCE OF THE HOMINIDAE

In Table 6.1 are listed 11 important events bearing on the emergence of the Hominidae as the result of a series of evolutionary divergences. Items 1 through 3 are based largely on molecular evidence; items 4 and 5, on geological evidence; items 6 through 9, on paleontological and paleobotanical data; and items 10 and 11, on the anatomy of the presumptive emergent hominid fossils.

Because the molecular evidence has played the most significant role in determining the time of hominid emergence, it is worthwhile to consider it here in some detail.

At one time, the gross anatomy of organisms, present and past, enlivened by insights from comparative anatomy, gave practically the entire ground for statements about

TABLE 6.1	Hominoid Divergences and the Emergence of the Hominidae	
		Approximate Age (Myr B.P.)
1.	Divergence of orangutan lineage from Hominoidea	17.3–14.0
2.	Divergence of gorilla from other African hominoids*	9.0–5.9
3.	Chimpanzee-hominid divergence, inferred appearance of Hominidae*	6.4–4.9
4.	Uplift, cooling, and aridification of Africa	7.0–5.0
5.	"Messinian crisis," the drying up of the Mediterranean	6.0–5.0
6.	Spread of African savannah	6.0–5.0
7.	Retreat of African tropical and subtropical forest	6.0–5.0
8.	Afro-Asian faunal interchange	6.0–5.0
9.	Many African faunal changes	6.0–5.0
10.	Earliest known fossils identifiable as very probably hominid	5.5–4.0
11.	Earliest fossil evidence of hominid bipedalism	4.0–3.5

*Some investigators believe the evidence indicates that gorilla, chimpanzee, and the hominids diverged at virtually the same time, between 7 and 5 Myr B.P.

biological affinities and descent. On this basis, Simpson (1945) placed all apes in one family, the **Pongidae,** and referred past and present members of the genus *Homo*, as well as *Pithecanthropus* (now *Homo erectus*) and *Eoanthropus* (the Piltdown remains, later exposed as forged), to a separate family, the Hominidae.

The enormous authority of Simpson's classification, and his supporting argument based on his definition of the principles of taxonomy, ensured that apes and hominids were kept apart in two different families by most scholars for a long time afterward. This position became entrenched with the publication of LeGros Clark's *The Fossil Evidence for Human Evolution* (1955, 1964). Clark offered clear and extensive zoological definitions of the two families Hominidae and Pongidae. Although early members of the two families were more alike than their later, derived descendants, including those living today, it was still possible for Clark to define the morphological trends characterizing the two families in such a manner that even early fossil hominoids could be assigned to one family or the other.

Thus the formal position of hominid and ape systematics, for about the last half century, has been as follows:

ORDER	Primates
SUBORDER	Anthropoidea
SUPERFAMILY	Hominoidea
FAMILIES	Hominidae, Pongidae

This classification of hominids and pongids is based on the structural affinities and differences between the present-day and earlier forms included in the two families.

Since the early 1960s, hominid origins have engaged the attention of molecular biologists. Their approach has been founded, in the main, on the biochemical makeup of living organisms. These studies have shown that there is a far closer biomolecular affinity between humans and the African great apes than we might have predicted from their comparative anatomy alone. Indeed, at the molecular level, gorilla, chimpanzee, and human beings are more closely related to one another than is any one of these to the Asian great ape, the orangutan. The molecular data are thus forcing a taxonomic realignment of the hominoids.

If, at the genetic level, the African apes are closer to humans than they are to the orang, it makes little sense to continue referring all of the great apes, Asian and African, to one family and assigning humans to another. It would seem more logical to place humans and the African great apes in a single family (for which the name Hominidae may be suitable), and to refer the Asian great ape, the orang, to a separate family, the Pongidae. Such a resorting of the hominoids has been proposed by several scholars (from Goodman, 1974, onward).

The classification of hominoids is therefore in a state of flux. The comfortable Simpson-LeGros Clark sorting has been assailed. The challenge has come not only from the differing data of paleontologists and molecular biologists, but also from the emergence of two major competing philosophies of taxonomy: Evolutionary systematics and cladistics. **Evolutionary systematists** take into account the unique morphological and behavioral features of *Homo* that distinguish it from the great apes: Thus they classify *Homo*, its ancestors, and its collaterals, in the Hominidae; the great apes, both African and Asian, are referred to the Pongidae. On the other hand, the **cladists,** following the work of Hennig (1950, 1966), classify groups according to the sequence of divergences, or branchings, into **sister groups:** As shown in Figure 6.1, because the evidence points to an ancient branching time for the orang from the African hominoids, cladists would place the orang in one family, the Pongidae, and its sister group, comprising humans, chimpanzee, and gorilla, in the Hominidae (Conroy, 1990; Mayr, 1981).

A new systematics is needed, one consistent with both the fossil evidence and the molecular interpretations and able to satisfy both the evolutionary systematists and the

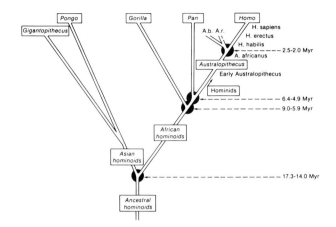

FIGURE 6.1

Principal divergences in the evolution of the Hominoidea. The estimated ages for the divergences leading to *Pongo* (orangutan), *Gorilla*, and *Pan* (chimpanzee) are based on molecular biological data. The estimated age of the most recent divergence—that leading to "A.b." *(Australopithecus boisei)*, "A.r." *(Australopithecus robustus)*, and *Homo habilis*—is based on geologic dating of relevant fossil material. Molecular evidence is accepted here as indicating that the divergence of *Gorilla* from African hominoids occurred somewhat earlier than that giving rise to *Pan* and hominids. However, some molecular biologists consider that the evidence points to a virtually simultaneous origin of *Gorilla*, *Pan*, and the Hominidae (in the narrow sense). The diagram shows a threefold splitting (trifurcation) of the hominid lineage at 2.5 to 2.0 Myr B.P. to give rise to *A. boisei*, *A. robustus*, and *H. habilis*. It is possible, however, that an earlier split may have generated *A. boisei*, and that a somewhat later split led to *A. robustus* and *H. habilis*. Further research and the discovery of more fossil specimens are necessary to resolve these differing interpretations.

cladists. In the meantime, in the absence of such a consensus, the term hominid is used here to apply exclusively to the family of humans and their Plio-Pleistocene ancestors and collaterals.

The divergence times listed in Table 6.1 show that the last African hominoid divergence was that between protochimpanzees and protohumans (Figure 6.1). This divergence is estimated to have occurred between 6.4 and 4.9 Myr B.P., that is, during approximately the last one million years of the Miocene Epoch (Table 6.2). From that time onward, we may say that hominids, in the narrow sense of the term, were in existence, or at least that the molecular genetical composition of hominids was existent. Of course, this molecular evidence cannot reveal more than simply the molecular makeup of the protohumans; that is, it cannot tell us whether they stood upright habitually, walked on two feet (bipedally), had small canine teeth, or possessed enlarged brains, as later hominids did. Not the molecules, but only the fossils can answer *these* queries.

A handful of fossil hominoid bones is all we possess to document directly the time of the inferred chimpanzee-human divergence. Of these, a few fragments may represent the pre-divergence common ancestors; one or two, the post-divergence protochimpanzees; and perhaps a couple, the post-divergence protohumans, that is, the earliest hominids. The available evidence is appallingly skimpy, and excavators of the past badly need to probe this time—between about 7.0 and 4.0 Myr B.P.—in order to fill in these gaps in the fossil record. In other words, we need direct fossil evidence to confirm the inferences drawn from the molecular data. Only one fossil specimen, a fragment of a thigh bone (femur) of a 16- or 17-year-old adolescent, recovered from Maca in the middle Awash Valley of Ethiopia and dated to be between 4.0 and 3.5 Myr in age, shows anatomical features that appear to point to its owner having walked bipedally. If this claim is correct, this specimen provides the

TABLE 6.2   Recent Stages in the Earth's History and the Time of Mankind

ERA	PERIOD	SYSTEM	EPOCH	Subdivision of EPOCH	Millions of Years B.P.	Time of Mankind
Cenozoic from 65 Myr B.P. to the present	Quaternary	Quaternary	Holocene		0.01	
			Pleistocene	Upper	0.125	
				Middle	0.7	
				Lower	1.8	
	Tertiary from 65 Myr B.P. to the present	Neogene	Pliocene		5.0	
			Miocene		25.0	
		Paleogene from 65 to 25 Myr B.P.				

earliest material evidence of hominid bipedalism (Figure 6.2). As such, it would indicate the presence in Africa of upright-walking hominids during the middle of the Pliocene Epoch, just less than 4 Myr B.P. (Table 6.2).

If we accept the evidence of the Maca femur, it may be stated that whereas "molecular hominids" were in existence between 6.5 and 5.0 Myr B.P., "morphological hominids" (as defined by bipedalism) are known *at least* since 4.0 to 3.5 Myr B.P.

It would be quite wrong to think that even excellent evidence, available from some rich fossil trove at just the right depth, would necessarily show that the molecular constitution and the anatomical makeup which distinguish hominids from apes appeared simultaneously. There are many features differentiating the various modern hominids (Table 6.3).

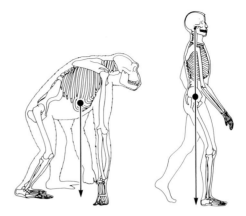

FIGURE 6.2

Skeletons of gorilla (left) and modern man (right). In an anthropoid ape, such as the gorilla, the stance and gait are generally in the oblique quadrupedal position and the weight line falls between the forelimbs and the hind limbs. In upright standing and walking man, the axis of the body mass (the "center of gravity") passes from the joint on the base of the cranium (close to the vertebral column), through the hip joints on either side, to the feet. Although various forms of bipedalism occur in primates other than man, the peculiarly human form of bipedalism is a distinctive adaptation that was acquired early in the process of hominization. Its acquisition has been accompanied (and indeed has been made possible) by anatomical adjustments affecting every part of the skeleton and the locomotor apparatus from the cranial base to the feet.

From the fossil record, it is clear that some of these traits occurred in their modern human state relatively early in the lineage leading directly to *H. sapiens*, and others only later. This pattern of change, with different characters evolving at differing rates, is known as **mosaic evolution** (Figure 6.3). It would be reasonable to assume that mosaic development applied also at the time of emergence of the Hominidae. At that stage, perhaps mosaicism was expressed not only among anatomical features, but also among the various sets of molecular indicators of hominid status, as well as between molecular markers, on the one hand, and morphological indicators, on the other.

PALEOECOLOGY OF HOMINID EMERGENCE

"No man is an island, entire of itself," wrote John Donne; "every man is a piece of the continent, a part of the main." So it was with man's emergence: It happened on a continent busy with change.

What was Africa like when the hominids emerged? In the Late Miocene, that is, the last few millions of years before 5.0 Myr B.P., Africa was characterized by a cooling and increasingly arid climate. It was the time of the **Messinian crisis,** when desiccation of the Mediterranean Basin permitted a major interchange of land fauna to occur between Africa and Asia. At this specific juncture, the molecular evidence suggests that the Hominidae first made their appearance, and Brain (1981) opines that the emergence of the hominids might have been a consequence of this cooling event.

The Late Miocene was an era of dramatic change among the flora and fauna of Africa (Tobias, 1984). A primitive member of the bear (ursid) family, *Agriotherium,* made its last appearance, prior to its vanishing from the African scene. New tribes of bovids became prominent, including several endemic tribes of typical savannah antelopes (Vrba, 1985a,b).

Members of the elephant family appeared in Africa, as did hyenas and saber-toothed cats. The pig family (Suidae) diversified greatly, and such rodents as *Otomys* became widespread. It is likely that some of these changes, especially those of some antelopes, elephants, carnivores, and rodents, flowed from the major faunal interchange accompanying the Messinian crisis. That is, species new to Africa moved onto the continent from western Asia. Other changes, particularly of some of the bovid tribes (Vrba, 1985b), seem to have occurred among groups already present in Africa, and to have been related especially to the Late Miocene cooling. Thus the African savannah fauna made its debut

---

TABLE 6.3   **Principal Features of Modern Humans Not Exhibited by Extant Nonhuman Primates**

A. Morphological

1. Habitual fully erect posture
2. Bipedal locomotion marked by striding gait and running
3. Lower limbs much longer than upper limbs
4. Comparatively vertical face
5. Great reduction in the projection of jaws
6. Great reduction of the canine teeth
7. Absence of a bony diastema in the upper jaw for the reception of the tip of the lower canine
8. Prominent nose with elongated tip
9. Median furrow (philtrum) of the upper lip
10. Outwardly rolled mucous membrane of the lips
11. Well-marked bony chin
12. A forward lumbar convexity ("hollow back")
13. Nonopposable great toe, set in line with other toes
14. Foot arched transversely, and from front to back
15. Relative hairlessness of body
16. Absence of tactile hairs
17. Brain much larger than largest nonhuman primate brain, both absolutely and relatively
18. Occiput of the cranium projects backward
19. Highly rolled margin (helix) of the ear
20. Absence of premaxillary bone from anterior aspect of face
21. Iliac blades (fossae) face each other
22. Longer postnatal growth period

B. Functional, Cultural, and Social

1. Articulate speech and language
2. Implements and other features of a complex culture
3. Dependence for survival primarily on cultural adjustment
4. Potentialities for development of intelligence
5. Capacity for symbolic thought: Abstract thinking, substitute activity

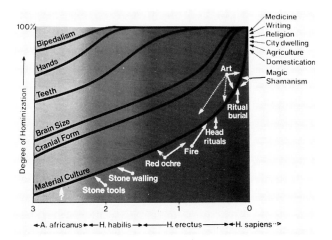

FIGURE 6.3

Diagrammatic representation of the mosaic pattern of hominid evolution; beginning with *Australopithecus africanus,* at the left, hominids are represented by four species grades, over a period of three million years, leading to *Homo sapiens* of the immediate past and the present. The plots portraying the development of various hominid physical traits reach the top of the diagram (representing 100% hominization, the attainment of modern human structure) at varying times, and they approach complete hominization at varying rates. Thus the fossil evidence indicates that modern hominid bipedalism was attained early, whereas modern human brain size was acquired late. The pathway of cultural evolution is shown in the lowermost plot on the diagram.

about that time. Among this fauna, it seems, walked the earliest hominids. Indeed, Vrba (1985a) suggests that the family of man was probably a "founder member" of the African savannah fauna.

Simultaneous with the advent of the savannah fauna, there is evidence of the spread of savannah flora in Africa. In the Cape Province of South Africa, for example, there are signs that the subtropical forest began to be replaced by "fynbos" shrubland cover. Forested and heavily wooded terrane gave way in many places to more lightly vegetated, open woodland and savannah. Hence, both the fauna and flora of Africa underwent drastic changes at the very time that the hominids are inferred to have emerged, namely, between about 7 and 5 Myr B.P. Clearly, the primates were not changing in a vacuum: All of Africa seethed with climatic shifts and biotic transitions and innovations; uplift of the continent, especially of its eastern part, heightened the tendency to cooling; animals new to Africa migrated onto the continent from the previously established Asian ("Palasiatic") fauna.

The living apes of Africa are found exclusively in the wet forest of the middle reaches of the continent. It is likely that ancestral apes, too, were forest-dwelling creatures (cf. Simons, 1985). The spread of lighter woodland and savannah, and the retreat of the margins of the primeval forests, could well have created conditions in which the tendency toward an erect posture and bipedalism was favored. The ability to run across the high grass cover of the savannah, perhaps from one woodland stream to another, might have held advantages for those apes that could "walk tall," since uprightness would have enabled its possessors to see over the high grass and to watch out for predators like lions and saber-toothed big cats. Seemingly, it was under such a set of conditions that the Hominidae made their appearance.

In short, the challenge resulting in the emergence of the hominids could well have emanated from the striking environmental modifications that took place during the Late Miocene. These circumstances were not confined to cooling and its direct effects on the last common ancestors of the hominids and the African great apes (Tobias, 1985). Rather, we have to deal with a more complex network of causal mechanisms. The underlying causes of

African cooling were global trends (Boaz and Burckle, 1984), Antarctic events (Flohn, 1984), and intra-African upheavals in the form of a somewhat earlier major continental uplift (Partridge, 1985). Each of these affected Africa and its biota secondarily, in diverse ways, such as through changes in the indigenous fauna and by immigration of new faunal elements; through geomorphic responses, such as the modification of drainage systems, and both the creation and the demise of lakes; through deforestation and the spread of more open, less protective floral conditions; and through other forms of ecological diversification. With new immigrant animals, freshly evolved endemic species, spreading savannah, and prevalent cooler, more arid conditions, environmental challenges to survival became more exacting.

From this African Late Miocene maelstrom, one of the continent's small-brained, little specialized ape populations came forth as tottering bipeds with emancipated hands and fingers: They were the founding parents of the family Hominidae, members of the genus *Australopithecus* (Figures 6.4–6.9). Audaciously, they slipped the familiar, comfortable constraints of their former forest retreat. With curiosity, enterprise, and opportunism, they ventured onto the plains of the high, inland plateau of Africa—and in time, they conquered the savannah.

The Last 1.5 Myr of the Pliocene

Whereas the known hominid record before 3.5 Myr is very sparse, between 3.5 and 1.8 Myr B.P. we have a large number of African hominid fossils (Figure 6.10). They are of various kinds and have thus been classified into different species—so much so that in one recent article it is claimed that as many as five hominid species coexisted by 2.0 Myr B.P. (Foley and Dunbar, 1989). Although not many investigators would go along with the proposition that the hominid world was so highly split at that time, most would recognize at least three different species as coexisting by the end of the Pliocene Epoch, namely *Australopithecus robustus*, *A. boisei*, and *Homo habilis*.

FIGURE 6.4

Facial view of the Taung skull, the type specimen of *Australopithecus africanus*, discovered in late 1924 at the Buxton Limeworks at Taung, South Africa, and recognized by R. A. Dart, in 1925, as representing a previously unknown kind of higher primate. Except for the first permanent molars, all of the erupted teeth are deciduous; thus, by modern hominoid standards, this skull would be that of an individual between three and five years of age. The Taung skull was the first of Africa's "ape-man" (australopithecine) fossils to be found; Dart's recognition of its hominid features constituted one of the most important events in the history of paleoanthropology in the 20th century.

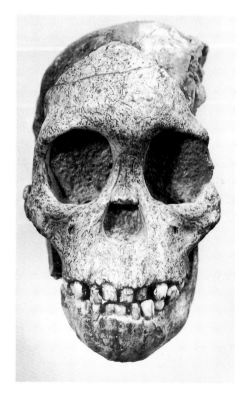

FIGURE 6.5

Right lateral view of a cranium of *Australopithecus africanus* (specimen Sts 5, "Mrs. Ples"), from Sterkfontein, Transvaal, South Africa. This splendidly preserved specimen was discovered by R. Broom and J. T. Robinson on 18 April 1947 in a stratum later designated by T. C. Partridge as Member 4 of the Sterkfontein Formation. Like that of other australopithecines, the cranial capacity (485 cm^3) of this specimen is small. Noteworthy is the degree of prognathism, the protrusion of the upper jaw. The teeth are missing from this otherwise magnificently well-preserved specimen (scale in centimeters).

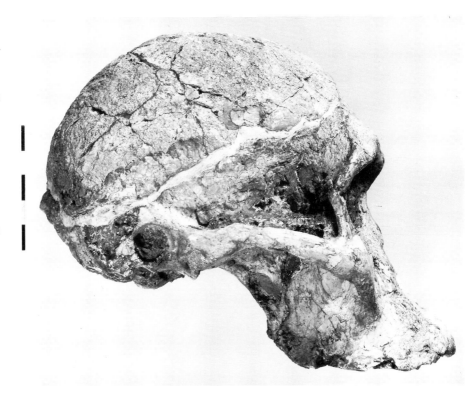

Thus, at some time between the emergence of the earliest hominids and the end of the Pliocene, there must have occurred one or more splittings of the hominid lineage. Recently obtained fossil evidence would place a major splitting at about 2.7 to 2.5 Myr B.P. (Figure 6.11). Thereafter, at any one time at least two different lines of hominids existed side by side. Because only one of those derivative lineages could have led to the genus *Homo* and, ultimately, to *H. sapiens*, it follows that the split brought into being some hominids that were *not* ancestral to modern man. When this fact was first realized, it came as somewhat of a surprise, because up to that time it had been widely and rather naively assumed that all earlier hominids were ancestral to the modern hominid, *H. sapiens*. Indeed, until fairly recently, some investigators were trying to fit all of the ancient hominids on the same

FIGURE 6.6

Posterior view of a cranium of a young adult of *Australopithecus africanus transvaalensis*, collected from Member 4 of the Makapansgat Cave Formation, about 200 miles north of Johannesburg, in the northern Transvaal, South Africa. Part of this specimen was found in December 1958 by J. Kitching, who discovered the remainder of the specimen in April 1959 (scale in centimeters).

FIGURE 6.7

A portion of the lower jaw of an adolescent *Australopithecus africanus,* found in 1948 by A. R. Hughes during systematic sorting of dumps at the Makapansgat Limeworks, northern Transvaal, South Africa. The photograph shows the mandible in the form in which it was discovered, partly embedded in a calcified cave deposit (breccia), prior to its excavation from the rock matrix (compare with Figure 6.8).

FIGURE 6.8

The same specimen of the lower jaw of *Australopithecus africanus* as that shown in Figure 6.7, after its excavation from the rock matrix. When the individual possessing this mandible died, at an age of 10 to 11 years, the second permanent molars had erupted, but not yet the second premolars or the wisdom teeth. The geological age of the specimen is about 3 Myr.

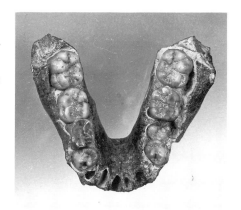

FIGURE 6.9

Left lateral view of the lower jawbone of an australopithecine, collected at Makapansgat, Transvaal, South Africa (scale in centimeters).

FIGURE 6.10

Map showing the most important African sites of early hominid discoveries. Five such sites are in South Africa; three are in Tanzania ("Garusi" marks also the famous footprint site of Laetoli); five are in Kenya; and three are in Ethiopia ("Afar-Awash" refers to the site commonly known as Hadar; the more recently discovered Maka site, in the middle Awash, is not shown). The East African sites are mostly open sites associated with lacustrine and riverine deposits of the Great Rift Valley. The South African sites occur in sealed dolomitic limestone cave deposits (those of the Transvaal), or in lime tufa deposits (the Taung site).

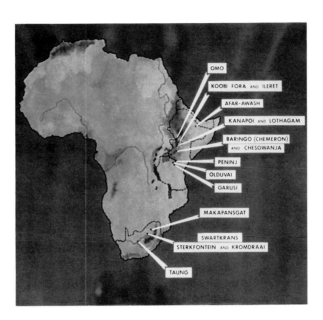

PROVISIONAL SCHEMA OF HOMINID PHYLOGENY

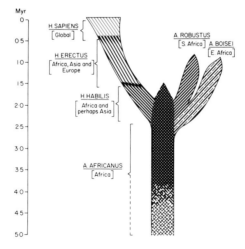

FIGURE 6.11

A provisional schema of hominid phylogeny based on the fossil record as known at the beginning of the 1990s. The evidence suggests that over a long period, from about 5.0 to 2.5 Myr B.P., the hominid family was represented by a single lineage that exhibited little morphological change (a lineage denoted here as *Australopithecus africanus*, the earlier stages of which are interpreted by some scholars as representing a different species, *A. afarensis*). About 2.6 to 2.5 Myr B.P., the lineage underwent a split (a cladogenetic event), or possibly two or three such splits, an evolutionary radiation that led to several derivative lineages. While some populations of late or "derived" *A. africanus* might have persisted for a time, showing relatively little further change, at least three lineages flowed from this evolutionary radiation: *A. boisei* (known from East Africa); *A. robustus* (known from South Africa); and *Homo habilis* (known from East and South Africa, and perhaps Asia).

lineage, by dismissing the variations among them as having resulted largely from sexual dimorphism! Now we know that things were decidedly more complicated.

As Table 6.4 indicates, the changes in the hominids during the later part of the Pliocene were played off against the background of another phase of cooling and drying in Africa, further opening of the woodland and spread of the savannah, and many faunal changes.

The earliest hominids of which evidence is known were small, lightly built, small-brained, bipedal organisms: At 3.5 to 2.8 Myr B.P. in East Africa, they have been designated *Australopithecus afarensis;* at 3.0 to 2.4 Myr B.P. in South Africa, they are named *A. africanus.* Many investigators regard these two groups of fossils as representing two good species; others have drawn attention to the marked resemblances between them, and consider the amount of difference between them as no more than would be expected of two subspecies of the same species (Figures 6.11 and 6.20).

At this stage in hominid evolution, from 3.5 to 2.8 Myr B.P., the known fossil hominids are not so diverse that we cannot regard them as representing earlier and slightly later populations of a single lineage. Thus *A. afarensis* and *A. africanus* may well represent relatively little changed members of the original hominid lineage that followed the chimpanzee-human divergence. This hypothesis may be tested by further discoveries of fossils representing the immediately post-divergence, protohuman population.

Several different hypotheses endeavor to portray the splitting phenomena and relationships of the hominids between 3.0 and 2.0 Myr B.P. The view which is here considered most likely is that some populations of *A. africanus,* after about 2.8 Myr B.P., underwent morphological changes into what Skelton and his co-workers (1986) called a "derived *A. africanus.*" Among these changes, the base of the cranium became much shortened, as compared with the elongated, apelike cranial base of earlier occurring specimens of *A. africanus;* the foramen magnum assumed a relatively more forward position on the cranial base (Figure 6.12); and the braincase (calvaria) underwent some broadening.

According to this view, these changes presaged the dramatic splitting of the hominid lineage. From the "derived *A. africanus*" there stemmed (1) the East African hyper-robust

TABLE 6.4	**Some Critical Events in Hominid Evolution and Ecology from 3.5 to 2.0 Millions of Years Before the Present**	
		Approximate Age Myr B.P.
1.	Further cooling and aridification of Africa	2.5–2.0
2.	Further opening of woodland and spread of savannah	2.5–2.0
3.	Many changes in mammalian fauna of Africa (baboons, elephants, pigs, bovids, hippopotami, saber-toothed cats, rodents)	2.5–2.0
4.	Hominid fossils known	3.5–2.0
5.	Differentiation of postulated "derived *A. africanus*"	2.8–2.5
6.	One or more splittings of hominid lineage	2.6–2.5
7.	Earliest known *Australopithecus boisei* fossils	2.5
8.	Earliest known *Homo habilis* fossils	2.2–2.0
9.	Earliest known stone cultural remains	2.5
10.	Acquisition of spoken language (as here inferred)	2.5–2.0
11.	Earliest modern human brain form	2.0
12.	Earliest signs of marked brain enlargement in hominids	2.0

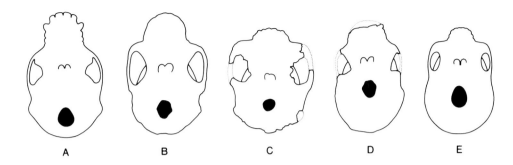

FIGURE 6.12

One of the adjustments that accompanied the evolution of bipedalism was a progressively more anterior placement of the foramen magnum (shown as black oval-shaped areas) and the flanking occipital condyles, by which the skull articulates with the vertebral column. (A) *Gorilla* (modern); (B) *Australopithecus africanus*, from Sterkfontein; (C) *A. boisei*, from Olduvai; (D) *Homo habilis*, from Olduvai; (E) *H. sapiens* (modern).

australopithecine, *A. boisei*; (2) the South African robust australopithecine, *A. robustus*; and (3) a new genus and species, *Homo habilis*. It is possible that (1), (2), and (3) all stemmed from a single splitting into three derivative lines; or that more than one split was involved, for example, (1) splitting from (2 + 3), followed by (2) and (3) splitting from each other, or (3) splitting from (1 + 2), followed by (1) and (2) separating from each other (Figure 6.13).

Whatever the precise pattern of evolutionary changes and splitting, between 2.5 and 2.0 Myr B.P. a new genus and species, which L. S. B. Leakey et al. (1964) called *Homo habilis*, had come into being (Figure 6.11). This species was regarded by its authors as the direct ancestor of the two subsequent species of *Homo*, namely, *H. erectus* and *H. sapiens*. At the same time, much differentiated species of *Australopithecus* had appeared. In contrast with *Homo*, these bipedal, robust and hyper-robust australopithecines were characterized by heavier bodies, enlarged cheek teeth (premolars and molars), and slightly larger brains. Judging from the markings on their teeth, they might have had a tough and abrasive diet

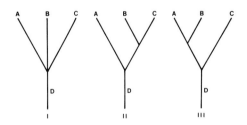

FIGURE 6.13

Three of the possible patterns of branching of an ancestral lineage of *Australopithecus africanus* (D) to give rise to *A. boisei* (A), *A. robustus* (B), and *Homo habilis* (C). In pattern I (left), the common ancestor split, at more or less the same time, into the three derivative lineages. In pattern II (middle), *A.* *boisei* (A) is postulated to have split off relatively early, after which a second split separated the two South African species, *A. robustus* (B) and *H. habilis* (C). Pattern III (right) suggests that the lineage of early *Homo* (C) split off relatively early, after which a second split separated the "hyper-robust" (A) and robust (B) australopithecines. At present, the available fossil record is neither sufficiently complete nor adequately dated to enable paleoanthropologists to reach a consensus as to which of these three patterns is more likely to be correct.

based on hard nuts and seeds (Grine, 1986; Walker, 1981). It appears that, under more arid and open environmental conditions, the robust and hyper-robust australopithecines adapted behaviorally, by a change of diet, facilitated structurally by a change of dentition and its supporting bony structures (Foley and Dunbar, 1989).

While *Australopithecus robustus* and *A. boisei* were adapting, in their way, to the new harsher conditions in Africa, they were also becoming specialized off the main line of human descent. On the other hand, *Homo habilis* was developing a different set of characteristics that placed it clearly on the lineage leading to modern mankind.

A crucial piece of evidence in the first recognition of *H. habilis* as a more "hominized" creature than *Australopithecus africanus* was provided with Tobias's (1964) demonstration of the larger endocranial capacity—and, by inference, of the larger brain size—of the primary "type" specimen of *H. habilis* (Olduvai hominid 7), with its extensive parietal bones. Since then, the validity of such endocranial capacity evidence has been verified by R. Holloway, D. Pilbeam, and P. V. Tobias for this and a number of other specimens assigned to *H. habilis*. Holloway (1975) provided new estimates of the endocranial capacities of *A. africanus* fossils from the Transvaal. As a result of his work, and of a newer study by Conroy et al. (1990), it has become possible to calculate a sample average of 440 cm^3 for six *A. africanus* crania. For *H. habilis*, in contrast, the average capacity for a sample of six specimens from East Africa is 640 cm^3, some 45% greater than the mean for *A. africanus* (Figure 6.14).

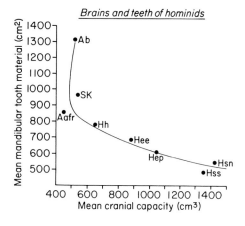

Brains and teeth of hominids

FIGURE 6.14

The average endocranial capacity and the average tooth size for each of a series of hominid species or subspecies (Ab = *A. boisei*, the "hyper-robust" australopithecine; SK = Swartkrans, *Australopithecus robustus crassidens*, a subspecies of *A. robustus* from Swartkrans, South Africa; Aafr = *A. africanus*; Hh = *Homo habilis*; Hee = *H. erectus erectus*, a subspecies of *H. erectus* from Java, Indonesia; Hep = *H. erectus pekinensis*, a subspecies of *H. erectus* from northern China; Hsn = *H. sapiens neanderthalensis*, Neandertal Man; Hss = *H. sapiens sapiens*, the modern subspecies of *H. sapiens*). The tooth size, the "mandibular tooth material," is calculated as the sum of the crown areas of the cheek teeth (those of the two premolars and the three molars) for each individual in each taxon; the mean tooth material for all the available individuals in each taxon is then calculated and plotted. Early occurring hominids lie on the left of the graph, and more recent hominids on the right. Broadly, the diagram shows that from an early presumptive ancestor, "Aafr" (*A. africanus*) that exhibited small endocranial capacity and moderate tooth size, two lines of change can be recognized. In one of these, shown by the almost vertical portion of the curve (at the upper left) leading from "Aafr" to "SK" (Swartkrans = *A. robustus* crassidens) to "Ab" (*A. boisei*), there occurred marked enlargement of teeth and a very small increase in brain size; this line became extinct between 1.5 and 1.0 Myr B.P. The other line of change, shown by the descending portion of the curve leading from "Aafr" to "Hh" (*Homo habilis*) and ultimately to "Hss" (*Homo sapiens sapiens*) and "Hsn" (*Homo sapiens neanderthalensis*), includes the various species and subspecies of *Homo*. This latter line is characterized by a marked increase in endocranial capacity and a progressive diminution in tooth size, culminating in the extraordinarily large brains and small teeth of *Homo sapiens*.

Brain size is obviously related to body size. Early in the 19th century, the French comparative anatomist Georges Cuvier first introduced the concept of **relative brain weight,** that is, the weight of the brain expressed as a fraction of the weight of the body. Later studies by O. Snell, E. Dubois, R. Bauchot, H. Stephan, and E. Thenius showed that a species in which the members are, on average, of heavier body weight, will be expected to have a larger average brain size than a species with lighter bodies. The endocranial capacity of *A. africanus,* in *absolute* terms, was no greater than that of modern anthropoid apes. However, from the evidence of the limb bones and the spinal column of these fossil hominids, it is probable that the body size of *A. africanus* was somewhat smaller than that of the extant apes. It follows, therefore, that the *relative* brain size of *Australopithecus* was slightly larger than that of modern apes. When several indices of brain development **(encephalization)** are employed to relate endocranial capacity to estimated body size, these show that the various australopithecine species were somewhat more encephalized than the modern chimpanzee (Holloway and Prost, 1982; Jerison, 1973; McHenry, 1982; Tobias, 1987).

To summarize, all of the various species in the genus *Australopithecus* have absolute and relative brain sizes which, on average, are slightly smaller than those of the living gorilla, but slightly larger than those of modern chimpanzees and orangutans.

Nevertheless, the notable hominid trend toward marked brain enlargement, both in absolute and relative terms, was not yet in evidence at the evolutionary stage represented by *Australopithecus.* Appreciable absolute and relative brain enlargement occurred only later, with the advent of the genus *Homo* (Figure 6.14): The postcranial bones of *H. habilis* clearly indicate that the members of this species were small-sized, virtually pygmoid in stature—and yet their brains were enlarged by 45% as compared with those of *A. africanus!* This was one of the most striking paleoanatomical features of *H. habilis.*

Of course, *Homo habilis* exhibited intraspecific variability, and among hominids, brain size is known to vary with age, with sex, and among individuals. Thus six crania attributed to *H. habilis* (four from Olduvai, in Tanzania, and two from East Turkana, in Kenya) had endocranial capacity "adult values," ranging from 510 cm^3 to 752 cm^3. The average for three specimens inferred to represent males of *H. habilis* was estimated by Tobias (1990a) as 688 cm^3 and for three putative females to be 592 cm^3. The index of sexual difference between these two small samples (13.9% of the male mean) was found to be comparable to index values typical of modern large hominoids, which range from 10.6% to 14.8% (Tobias, 1990a).

A second major group of features distinguishing *H. habilis* from *A. africanus* and modern great apes relates to the form of the brain cast. In *H. habilis* certain areas of the cerebrum are selectively enlarged as compared with others. The lower parts of the frontal lobe, and of the parietal lobe of the cerebrum, are enlarged in *H. habilis* as compared with these regions in endocasts of *A. africanus* and in the brains of modern great apes. These two regions are the seats, respectively, of Broca's area and of Wernicke's area, the two most important cortical speech areas. This evidence, supplemented by cultural inferences, led Tobias (1980, 1981, 1983) to suggest that *H. habilis* not only possessed the neurological basis of speech but, as evidently indicated by its relatively complex material culture, it probably had a rudimentary form of speech, a view supported by Falk (1983) and by Eccles (1989). Such evidence—both neuroanatomical and cultural—has not been found in *A. africanus.*

A third set of traits characterizing *H. habilis* is related to its teeth. Overall, *H. habilis* teeth were smaller in absolute size than those of *A. africanus.* Moreover, the premolars and the first permanent molars were attenuated from their outer (cheek) aspect, toward their inner (tongue) aspect (i.e., from buccal to lingual). Thus the shape of the tooth crowns appeared narrower and more elongated than the bulbously broad cheek teeth of the australopithecines. Differential reduction in tooth size along the tooth row also occurred in *H. habilis,* this reduction being apparently correlated with a different pattern of occlusion and attrition between the upper and lower teeth, as compared with the pattern seen in apes and in *Australopithecus*.

In addition, the nature of the base of *H. habilis* crania shows that the head of this species was carried in a better balanced fashion, upon a more upright vertebral column, than in *A. africanus*. That is to say, in *H. habilis*, anatomical adjustments to an upright posture and bipedal gait had been carried relatively further in the direction of the skeletal adaptations found in modern humans.

In short, *H. habilis* was nearer to modern humans than *A. africanus* in respect to the size and form of the brain, the size and morphology of the teeth, and the adjustment to an erect posture, as well as in many other features that are not detailed here.

THE DAWN OF STONE CULTURE: ETHOLOGY AND ECOLOGY

Within the last part of the Pliocene Epoch, perhaps as early as about 2.5 Myr B.P., evidence of a new dimension in hominid evolution appears in the records—the earliest stone tools (Figure 6.3).

For many years, some investigators (including the author) perceived what they took to be a correlation between the age of the earliest stone tools and the first appearance of *Homo habilis* (Leakey et al., 1964; Leakey, M.D., 1967, 1971; Tobias, 1971). Napier and Tobias (1964) used the association between *H. habilis* remains and early Oldowan tools as behavioral (**ethological**) evidence supporting the then disputed assignment of the *H. habilis* bones to the genus *Homo*. At that time, there was no good evidence for the occurrence of stone tools earlier than the oldest fossil bones of *H. habilis*. Although there seems good reason to believe that *H. habilis* was a confirmed toolmaker and a culture-bound hominid, the question remains whether this species was the earliest hominid taxon to display stone toolmaking. Since 1976, an increasing body of evidence has pointed to the occurrence of hominid implement-related activities earlier than the oldest records of *H. habilis*.

The oldest skeletal remains attributed to *H. habilis* date from about 2.2 to 2.0 Myr B.P. (Boaz, 1979). These are the habiline remains from Omo, in the south of Ethiopia. However, the oldest recorded occurrence of fossils of a species does not necessarily signify the date at which the species actually appeared; for *H. habilis*, therefore, that date may need to be set at least a little earlier than the 2.2 to 2.0 Myr B.P. indicated by the direct fossil evidence.

Although it is not long since claims for the existence of stone tools older than 2.0 Myr B.P. were seriously disputed, it has become clear that, in Ethiopia at least, some occurrences of stone tools are appreciably older. The most ancient traces of stone artifacts have come from the Gona River area, west of the main hominid-bearing locations of Hadar, Ethiopia. There, in the middle 1970s, Corvinus made the first discovery of stone artifacts in strata subsequently dated to have been deposited between 2.7 and 2.4 Myr B.P. (Corvinus, 1976; Corvinus and Roche, 1976; Taieb and Tiercelin, 1979). The age determinations were based on well-established techniques, namely, potassium-argon dating, fission track studies, paleomagnetic stratigraphy, and stratigraphic correlation (Harris, 1983; Roche and Tiercelin, 1977). Further artifacts were obtained by Roche, in 1976; Harris, in 1977; and by Harris and Taieb, in 1977. Harris (1983) has stated that these stone tools are similar to later assemblages known from Olduvai Bed I, in Tanzania, and from Koobi Fora, in northern Kenya.

If these tools have been dated correctly, they are certainly older than the earliest recorded members of *H. habilis*. Some slight revision of the earlier claimed dates for these implements may be necessary, perhaps to a maximum antiquity of 2.5 Myr B.P. (J. W. K. Harris, 1990, personal communication). If so, this would reduce the temporal priority of the earliest known tools over the oldest dated *H. habilis* remains to about 0.5 to 0.3 Myr. The argument that the earliest preserved *H. habilis* fossils have not yet been discovered may apply equally, of course, to the earliest stone implements so far recognized. Moreover, the margin of difference between the two lines of evidence is moderate, when we take into account the limits of precision of the dating methods used. Yet, the available dates suggest that toolmaking first appeared *prior to* (and *not* coincident with, or subsequent to) the branching of the hominid lineage that generated, inter alia, *H. habilis*. If this interpretation is correct, the development of stone toolmaking must then have occurred in some hominid populations before the emergence of *H. habilis*. The most likely candidate as fabricator of the oldest known stone tools would appear to be *A. africanus*, more particularly the "derived *A. africanus*" postulated by Skelton et al. (1986), a taxon whose members have thus been nominated as the earliest stone toolmakers (Tobias, 1989, 1990a, 1990b). The development of stone culture in advanced populations of *A. africanus* might have been a behavioral counterpart of the morphological changes distinguishing the "derived *A. africanus*" from the less specialized, earlier *A. africanus*.

According to this evolutionary scenario, stone cultural activities would have provided a critical element in the environment of the "derived *A. africanus*." The emergence of stone culture should thus be seen as part of the network of causal events that led up to the splitting of the hominid lineage. It may be suggested that the acquisition of stone implemental activities by some populations of the "derived *A. africanus*," but not by others, might have provided an evolutionary "bottlenecking mechanism" which facilitated branching and speciation. Seen in this light, culture—as reflected initially by evidence of stone toolmaking—played an important, probably a crucial, role in the genesis of *Homo* and of its earliest well-attested species, *H. habilis*.

After the major split (or splits) in the hominid lineage at about 2.6 to 2.5 Myr B.P. (Figure 6.11), the toolmaking habit would have persisted as a feature of substantial selective advantage in one, or both, or all of the derived lineages. We cannot deny toolmaking capacity to *A. robustus* and *A. boisei* after the split, and it has been suggested that there may be evidence at Swartkrans, South Africa, for such activities among *A. robustus crassidens* (Brain et al., 1988). However, it is the *Homo* lineage which seems to have made stone culture peculiarly its own. *H. habilis*, it seems, became **obligate stone toolmakers and users,** whereas the late robust and hyper-robust australopithecines might have been only **facultative toolmakers and users** (able to make and use stone tools, but equally able to survive without them). Following the inferences by Walker (1981), and Grine (1986), that *A. boisei* and *A. robustus* had a tough, abrasive diet, Foley and Dunbar (1989) suggest that the more arid and open environment, in which it is believed these late australopithecines lived, together with the greater amount of time they would have spent on feeding, would have created conditions unlikely to be ideal for the evolution of culture. If this reasoning were correct, it might explain why the robust and hyper-robust australopithecines, although inheriting the propensity for stone toolmaking activities, did not become culture bound. However, a difficulty with Foley's and Dunbar's interesting ecological analysis is that we have no evidence that these late australopithecines lived in a more hostile physical environment than did the early members of the genus *Homo*. Indeed, the remains of the two groups of hominids are found cheek by jowl with one another at Omo, in Ethiopia; Koobi Fora, in Kenya; Olduvai, in Tanzania; and at Swartkrans, in the Transvaal, South Africa. In other words, everywhere that they are known to have occurred, the late australopithecines and early *Homo* are **sympatric,** sharing the same terrane and environment. What seems to have differed between them was that they adopted different strategies for coping with the rigors of the environment. One such difference, it has been suggested, lay in the mode of preparation of food: The robust and hyper-robust australopithecines prepared their food predominantly in their mouths, chewing it between their greatly expanded upper and lower cheek teeth, whereas *H. habilis* prepared its food with hand-held stone tools prior to eating (Wallace, 1972, 1978).

From the time of appearance of *H. habilis* toward the end of the Pliocene Epoch, the archeological evidence indicates that the cultural dimension came to play a prominent part in the pattern of hominid life. Although the material cultural aspects of this pattern have been stressed here, an equally essential component of human culture is its *cognitive* underpinning (Lowrie, 1937; Washburn and Benedict, 1979). As the cultural component proliferated and diversified, with an inevitable deepening and widening of its cognitive basis, a considerable change in human adaptation must have occurred. Prior to the emergence of human cultural behavior, hominid ecological adjustments must have been essentially biological and social in character. The acquisition of a cultural dimension would have added appreciably to the range of possible mechanisms of adaptation available to the evolving human organisms.

Critical cultural elements listed in Table 6.4 relate to the acquisition of spoken language and to the emergence of modern human culture, but culture must have played a major role also in the later Early to Middle Pleistocene migrations of mankind, from Africa to Asia and to Europe. Culture might even have played a part in hastening the extinction of the last of Africa's australopithecines during the Early Pleistocene, about, or more recently than, 1.2 Myr B.P. (Klein, 1988; Tobias, 1986).

In the further evolution of mankind, biological and cultural events were concomitants. Indeed, it is often difficult to extricate the effects of one from those of the other. This dovetailing of biological and cultural mechanisms in human adaptation has persisted to the present. The survival of the earliest hominids seems to have been predicated essentially on biological factors. The survival and adaptation of the later Plio-Pleistocene hominids seem to have depended increasingly on cultural determinants, and many modern human populations have become largely, or even totally, culture dependent in their survival strategies.

In short, there has been a change in the major emphasis during human evolution from mainly biological mechanisms of adaptation at earlier stages, to predominantly cultural means at later stages. During this transition, spoken language has, from about 2 Myr B.P., played an increasingly important role as a means for the transmitting of the cultural and social heritage. The earlier "animal hominid" has been transformed into the later "human hominid." By the Early Pleistocene (1.8 to 0.7 Myr B.P.), humanity was becoming language bound and culture dependent. Armed with these qualities, the hominids broke the geographical bounds of their African origins and moved into Asia and Europe.

MAJOR EVENTS IN HOMINID EVOLUTION IN THE PLEISTOCENE EPOCH

As the world moved from Pliocene into the Pleistocene at 1.8 Myr B.P., hominids were still confined to Africa. They were represented by *H. habilis*, the toolmaking and probably speaking primate, and by the late-surviving robust and hyper-robust australopithecines. In the ensuing 1.8 million years, a number of important events occurred in the evolving hominids (Table 6.5).

Geographically, we find the hominids of this period bursting the bounds of their African origins and penetrating into southern Europe and eastward into Indonesia and then China.

Taxonomically, *H. habilis* gives way to *H. erectus* about 1.6 Myr B.P., and the latter, in turn, is supplanted by *H. sapiens* at a time variously estimated, but commonly set at about 0.5 Myr B.P. These three species appear to be sequentially evolving **chronospecies** of a single lineage, although some investigators, who believe that speciation occurs only, or predominantly, as a result of branching events, have felt it necessary to postulate two or more evolutionary branchings along this pathway.

Morphologically, we find evidence, inter alia, of further increases in absolute and relative brain size, the attainment of a brain form and pattern of convolutions indistinguish-

TABLE 6.5	Some Critical Events in Hominid Evolution in the Pleistocene Epoch	
		Approximate Age Myr B.P.
	1. Movement of hominids from Africa to Asia and Europe	2.0–1.5
	2. Emergence of *Homo erectus*	1.7–1.6
	3. Extinction of robust and hyper-robust australopithecines	1.2
	4. Acquisition of control of fire by *H. erectus*	1.3
	5. Emergence of *Homo sapiens* (variable, according to varied views on the nature, timing, and indeed the very existence of an *erectus-sapiens* transition)	0.5 (0.75–0.25)
	6. Earliest known "anatomically modern *Homo sapiens*" in Africa	0.125
	7. Earliest known "anatomically modern *Homo sapiens*" outside Africa	0.1
	8. Emergence of "modern human culture" (variable, according to one's concept of modern human culture)	0.1–0.025
	9. Earliest burial of the dead	0.1
	10. Earliest rock art	0.04–0.03
	11. Earliest protowriting (ideographs, pictographs, semasiographs)	0.03
	12. Earliest writing	0.005

able from those of modern humans, and further diminution of the teeth and of their supporting bony apparatus.

Culturally, Homo species become increasingly more dependent on cultural means of survival. This is reflected by the invariant association of their fossil remains with stone tools and similar objects and, beginning at about 1.3 Myr B.P., by evidence of their ability to control fire. In later stages, evidence of such advanced human behavior as ritual, burial of the dead, and of artistic expression is reflected in the record. Ultimately, as Marshack (1976) has shown, forms of protowriting (variously dubbed pictographs, ideographs, and semasiographs) anticipate the discovery in several parts of the world of one of modern humans' greatest gifts, writing.

HOMO ERECTUS AND *HOMO SAPIENS:* THE PATHWAY TO MODERN MAN

A Note on *Homo erectus*

One of the extinct species belonging to the family Hominidae was *Homo erectus* (formerly called *Pithecanthropus*). Members of this species differed in a number of respects from modern humans, which have been classified as a distinct primate species, *H. sapiens.* However, a number of fossil crania show features intermediate between those of *H. erectus* and *H. sapiens;* this and other factors have led some scholars to doubt whether it is justified to maintain the status of these taxa as two separate species.

Remains of *H. erectus* (at present, still recognized taxonomically by most investigators) have been found over large parts of the Old World. The first of these discoveries were centered in Asia. Similar fossils were found at Trinil, Kedung Brubus, Mojokerto, Sangiran, and Sambungmachan, in Indonesia (Figure 6.15). Another series of finds was made in China, especially in the cave of Zhoukoudian, west of Beijing, although virtually all of the remains from that locality were lost in 1941 during the Sino-Japanese War (Figure 6.16). After World War II, new discoveries of *H. erectus* were made in the same cave, as well as in two additional sites, in Lantian County, Shaanxi Province (namely, the Gongwangling and Chenchiawo sites), and at Longtandong, in Hexian County, Anhui Province.

By the end of World War II, the concentration of these early discoveries in Indonesia and China seemed to suggest that *H. erectus* was a peculiarly Asian expression of early mankind. Subsequently, new discoveries in Africa began to change this view, and it came to be realized that Europe, too, might have harbored *H. erectus.*

The African discoveries of specimens assigned to *H. erectus* came from Ternifine, east of Mascara, in Algeria; from Sidi'abd ar-Rahman, Thomas Quarry, Salé, and Rabat, in Morocco; possibly from Yayo (Koro Toro), in the Chad Republic; from Omo and Melka Kontouré, in Ethiopia; from East Turkana and possibly from Baringo, in Kenya; from

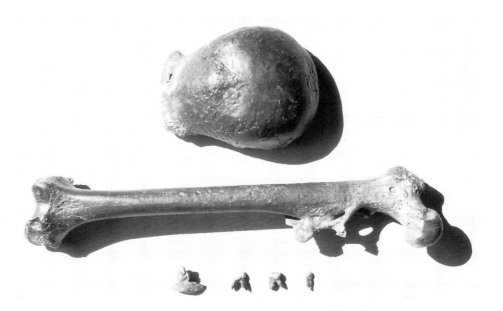

FIGURE 6.15

Skeletal remains of *Homo erectus* from Java, Indonesia. *Above:* The calvaria (braincase) found at Trinil on the Solo River by Eugene Dubois in October 1891; pear-shaped, heavily browed, medium-brained, and flattened above, this was the type specimen of a species originally designated by Dubois in 1894 as *Pithecanthropus erectus* (a taxon subsequently

renamed *Homo erectus erectus*). *Middle:* The completely intact thigh bone found by Dubois at Trinil in 1892. It was this typically modern femur that led Dubois to select "erectus" (upright) as the species name; interesting signs of pathology are retained near the upper end (at the right) of the bone. *Below (specimen at left):* The small remnant of a mandible from

Kedungbrubus, East Java, found by Dubois in 1890 and demonstrated by P. V. Tobias in 1966 to have belonged to a juvenile *H. erectus.* *Below (three specimens at center and right):* Two upper molars and a lower premolar (far right) ascribed to *H. erectus,* found by Dubois and one of his collectors between 1891 and 1898.

FIGURE 6.16

Restoration by Franz Weidenreich of the Chinese representative of *Homo erectus,* so-called Peking Man. Found in the cave of Zhoukoudian, near Beijing, this form of hominid was originally designated *Sinanthropus* by Davidson Black; later, it was "lumped" into *H. erectus* as a separate subspecies, *H. erectus pekinensis.* The Chinese subspecies was somewhat larger brained, smaller toothed, and later in time than the Javanese subspecies shown in Figure 6.15.

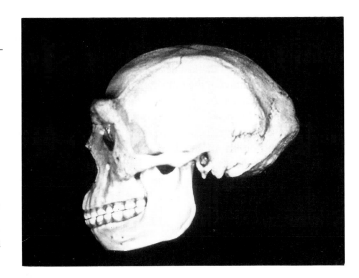

Olduvai Gorge, in Tanzania (Figure 6.17); and, probably, from Swartkrans, in South Africa.

Some European fossils were regarded by many scientists as belonging to *H. erectus,* although other investigators considered them to belong to "emergent *H. sapiens* (subspecies indeterminate)" (Howell, 1978). These specimens included a mandible, from Mauer, and cranial parts, from Bilzingsleben (both in Germany); an occipital bone and isolated teeth from Vertesszollos (in Hungary); a cranium from Petralona (in northeastern Greece); and fossils from Arago, Montmaurin, and perhaps from Lazaret (in France). Because the morphology of these European fossils differed in some respects from that of the Asian *H. erectus,* Howell considered them as early manifestations of *H. sapiens.* Others considered that these morphologic differences simply widened the known range of intraspecific variation in *H. erectus,* and that the fossils perhaps represented one or more geographical

FIGURE 6.17

The cranium of an African *Homo erectus* (hominid 9 from Olduvai Gorge in northern Tanzania), found by L. S. B. Leakey in December 1960. Note the very prominent browridge (at left); the retreating forehead; the very thick cranial bone; and, at the back of the cranium (at right), the marked nuchal ridge. The endocranial capacity of this specimen is about 1,000 cm^3, approximately three-quarters of the average capacity (viz., 1350 cm^3) of crania of modern *H. sapiens.* Although most investigators regard this specimen as representing a member of an East African subspecies of *H. erectus,* G. Heberer proposed in 1963 that the hominid represented be designated *H. leakeyi.*

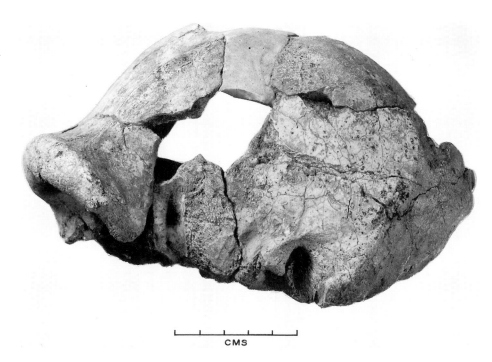

CMS

variants or subspecies of *H. erectus*. Indeed, several such subspecies have already been recognized—one from Java, one from China, one from northwestern Africa, and one from East Africa.

By the onset of the last decade of the 20th century, there was no unanimity on the exact pattern and timing of the transition from *H. erectus* to *H. sapiens*. Nearly all scholars in the field accepted that *H. erectus* was the predecessor of *H. sapiens*, but there was difference of opinion as to the classification of certain fossils, especially the European ones. Such differences were generally the result of varying definitions of the two species, varying judgments on the implications of the morphology, and varying concepts of the variability within each species. Where the fossils showed intermediate or transitional features between the two species, the problem was compounded, because the phyletic, evolutionary interpretation implied that some populations of the earlier species changed into the earliest members of the later species, and continuity between the two successive sets of populations was therefore to be expected. Where then would we draw the line between the two species?

Ironically, where the available fossil evidence was most abundant, the problem became especially acute methodologically. The more fossils there were on hand, the more intermediates were likely to be included and the greater the apparent continuity from one species to the other. This observed fact seemed to confirm that one form had evolved into another; at the same time, it made the problem of the classifier, the systematist, more difficult. This is a much less severe problem among living, contemporary species, the boundaries between which are usually clear-cut. Such boundaries, however, are less well defined among extinct species, especially where the evidence indicates a succession of species, broadly consecutive in time, in a single lineage such as the *H. habilis, H. erectus,* and *H. sapiens* succession. In such a sequence, it was not surprising to find specimens that showed intermediate characters between *H. habilis* and *H. erectus* (such as specimen SK 847 from the Swartkrans site), and others that appeared to be transitional between *H. erectus* and *H. sapiens* (such as the fossils from Petralona and Arago).

The various fossils most commonly accepted as belonging to *H. erectus* occur in strata of the Lower and Middle Pleistocene, dated as having been deposited in the period between 1.6 and about 0.5 Myr B.P.

All evidence points to *H. erectus* having possessed a complex culture, in advance of anything evinced in deposits containing remains of earlier hominids.

The Emergence of *Homo sapiens*

To this phase of hominid evolution, during the Middle Pleistocene, between about 0.7 and 0.125 Myr B.P., belong (1) a number of specimens that appear to have been transitional between *H. erectus* and early *H. sapiens;* (2) three important subspecies known as *H. sapiens soloensis, H. sapiens rhodesiensis,* and *H. sapiens neanderthalensis;* and (3) later subspecies of *H. sapiens,* the immediate precursors of living humans. Here, again, we are beset by problems of definition. At certain time levels, we possess virtually an *embarras de richesse,* so prolific is the known fossil record. During the later stages of this period, two factors heightened the chances of fossils being preserved, namely, the widespread adoption of cave dwelling by *H. sapiens* and the practice of burial of the dead. As a result, literally hundreds of fossils are known from this phase of human evolution. Indeed, the fossils are so numerous that it would be unhelpful to attempt to enumerate the individual specimens here.

By the last decade of the 20th century, the evolution of *H. sapiens* from *H. erectus* was very widely accepted, although substantial questions were unresolved. In particular, there remained uncertainty as to (1) a clear and widely acceptable definition of *H. sapiens;* (2) whether the origin of *H. sapiens* was a branching (**cladogenetic**) event, or was an example of phyletic speciation; (3) the dating of the earliest appearance of *H. sapiens* in the fossil record (see, for example, the wide range of dates noted in Table 6.5); (4) the geographical location, or locations, in which this species differentiated—or, indeed, whether it arose in only one area and then spread, or whether there were multiregional foci of origin, as C. S. Coon (1963) and some more recent investigators have held; and (5) the circumstances, climatic, demographic, and cultural, under which *H. sapiens* appeared.

Two useful definitions of *H. sapiens* appeared in the second half of the 20th century. The first, that of LeGros Clark (1964), comprised a list of essentially discrete traits. In the light of later studies on early *H. sapiens,* as Howell (1978) pointed out, this definition was so restrictive as to exclude "those antecedent populations of middle and earlier upper Pleistocene age which nonetheless have some features suggestive of the species" (Howell, 1978, p. 201).

The other useful formulation was that of Howell, who set forth a lengthy compilation of evolutionary trends and derived features, behavioral as well as morphological, which were apparent in Late Pleistocene to Recent populations of *H. sapiens sapiens,* when compared with *H. erectus.* For practical purposes, you may find the tabulation of morphological, functional, and behavioral traits in Table 6.3 of some help.

Despite the uncertainties just mentioned, the fossil record does include a number of fossil hominid remains from Africa, Asia, and Europe that appear to cast light on the origin of *H. sapiens.*

Some *H. erectus* to *H. sapiens* Intermediates or Transitional Forms

A group of fossils of this period from the Old World has long occasioned difficulties of diagnosis. Some experienced scholars regard them as members of *H. erectus,* and others as belonging to *H. sapiens.* The point of the matter is that they show transitional features between these two species.

In Europe, these fossils include crania from Swanscombe, England, and from Steinheim, Germany, as well as most of those encompassed by "European *H. erectus*" (such as the Mauer mandible and the Bilzingsleben cranium, from Germany; the Petralona cranium from Greece shown in Figure 6.18; the Vertesszollos remains from Hungary; and the La Chaise fossils from France). In general, these fossils are of a somewhat bigger brained and smaller toothed hominid than African and Asian *H. erectus,* and they show less striking development of such *H. erectus* traits as cranial thickening and exuberantly developed supraorbital and occipital bone ledges. Most of these specimens stem from the Middle Pleistocene and are dated to the period between 0.7 and 0.125 Myr B.P.

FIGURE 6.18

The Petralona cranium, found in 1959 in a cave on the Khalkidhiki Peninsula, near the village of Petralona, southeast of Thessaloniki, in Greece. The facial part is encrusted with the calcitic (stalagmitic) cave deposit from which the specimen was recovered. Originally identified as a Neandertal cranium, many investigators now regard this specimen as representing an intermediate or transitional form between *Homo erectus* and archaic *H. sapiens.*

In Africa, several specimens from Morocco in northwest Africa are thought to show *H. erectus* to *H. sapiens* transitional features. These include the Salé cranium and a specimen from Thomas Quarry. Although not precisely dated, both of these specimens appear to be derived from the Middle Pleistocene. Similarly, from East Africa, the Kibish II cranium from Omo, in southern Ethiopia; the Baringo mandible from Kenya; and, possibly, both the Ndutu cranium from Olduvai, Tanzania, and the Bodo cranium from Ethiopia, are all deemed to show intermediate features. These also are dated as being Middle Pleistocene, ranging up to the junction between Middle and Late Pleistocene. This array of African and European specimens may be supplemented by several finds of intermediate forms from China (for example, those reported from Maba, Changyang, Dingcun, Xujiayao, and Dali).

Collectively, this array of Old World hominid fossils suggests that, in the Middle Pleistocene, a number of departures from the *H. erectus* structure began to appear in some hominid populations. These anatomical departures were generally in a *sapiens* direction. The combined array of primitive, *H. erectus,* and derived, *H. sapiens,* characters in these populations could be regarded as marking the first emergence of the *H. sapiens* differentiation from the substrate of basically *H. erectus* populations.

The Earliest *H. sapiens* Populations

Apart from the *H. erectus* to *H. sapiens* transitional forms just referred to, other groups of early fossil hominids are recognized as constituting a full expression of *H. sapiens:* They are often spoken of as "archaic *H. sapiens*" (Stringer, 1974). Collectively, they show more robust skeletal features than anatomically modern *H. sapiens,* but they are generally less robust in bone structure than *H. erectus.* Although interpretations vary, at least three great subspecies of archaic *H. sapiens* are acknowledged by many scholars: *H. sapiens rhodesiensis,* of southern and eastern Africa; *H. sapiens soloensis,* of Java; and *Homo sapiens neanderthalensis,* of Europe, western Asia, and the Mediterranean littoral region. These three subspecies possess crania furnished with heavy supraorbital ridges and sturdy occipital ridges; the cranial vault bone is of moderate to marked thickness. Thus the three groups show features that reasonably are considered to have been derived from *H. erectus.* The average endocranial capacities in *rhodesiensis* (over 1250 cm^3) and *soloensis* (1151 cm^3) are greater than the average for any of the *H. erectus* samples, although not as great as that for modern *H. sapiens.* Hominid brains had clearly undergone further expansion as compared with *H. erectus.* As a result, the crania of these three subspecies show a lesser degree of upward tapering of the sidewalls of the braincase than in *H. erectus,* and a lesser degree of flattening above.

The *rhodesiensis* sample comprises two well-preserved crania, that from Kabwe (formerly Broken Hill), Zambia, and that from Hopefield (Elandsfontein), Saldanha Bay, South Africa. A lower jaw fragment from the Cave of Hearths, Makapansgat, northern Transvaal, has been allocated to the same subspecies, and some hominid cranial parts from Lake Eyasi, Tanzania, might also have belonged to *rhodesiensis.* Other archaic *H. sapiens* remains from Africa are that of Ngaloba from Laetoli, Tanzania, and, possibly, those of Bodo, Ethiopia, and of Ndutu, Tanzania.

The *soloensis* sample consists, in the main, of a clutch of crania from Ngandong in Java, Indonesia.

The subspecies *H. sapiens neanderthalensis* is the best known of the three subspecies of archaic *H. sapiens.* Specimens attributed to *neanderthalensis* are especially numerous in Europe and southwestern Asia, having been found in Belgium, France, Germany, Gibraltar, Yugoslavia, Italy, Portugal, Spain, the USSR, Israel, Iran, and Lebanon (Figure

FIGURE 6.19

A Neandertal cranium, from Monte Circeo, southeast of Rome (in Latina, Italy), found by A. Guattari in 1939. In comparison with *Homo erectus,* the crania of Neandertals have somewhat reduced, though appreciable, browridges; a well-developed "bun" at the rear end; marked elongation of the braincase (calvaria); and an enlarged endocranial capacity, often reaching values above the average of that in *H. sapiens* (see Figure 6.14).

6.19). At least half a dozen specimens recognized by some workers as Neandertal* have been discovered on the southern Mediterranean littoral, for example in Haua Fteah, Cyrenaica, and in Jebel Irhoud, Morocco, although the latter crania show more modern features than classical *neanderthalensis*. *H. sapiens neanderthalensis* lived from about 0.1 to 0.04 Myr B.P.

Well over 100 fossils from Europe, western Asia, and the Mediterranean littoral have been identified as belonging to the Neandertal group. The name Neandertal means, literally, "Neander valley," and stems from the first described specimen that was found in the valley of the Neander river in Germany, in 1856. At first, these remains were regarded as representative of a separate genus, as they were deemed to be "extremely primitive" and to differ markedly from modern human beings. No remains of earlier and less evolved hominids were then known. Later, the group was considered to be a separate species of the genus *Homo*. Finally, by the last quarter of the 20th century, Neandertal fossils were looked on as representing no more than an extinct subspecies of the species *H. sapiens*. Lately, a few investigators have tried to revive the idea that this group should be regarded as a separate species, *H. neanderthalensis*.

Neandertals were large brained, several specimens having endocranial capacities even bigger than the average for modern *H. sapiens*. Nevertheless, Neandertal crania retain some archaic features that appear to have been derived from an *erectus* ancestry. These features include a retreating brow; a prominent brow ridge; a cranial vault rather flattened on top; an occipital ridge; cranial bones that were a little thicker than those of modern *H. sapiens;* a poorly developed bony chin; a large broad palate and upper jaw; and a puffed out, inflated upper jawbone (maxilla). The postcranial skeleton was robust, each limb bone showing a thick cortex and large articular surfaces, while some specialized features also occur.

---

*The old spelling for the German word for valley was "thal," but for most of the 20th century, modern German orthography has spelled it "tal." The formal latinized species or subspecies name was thus originally spelled "neanderthalensis" and, under the rules of nomenclature, this spelling remains in force today. However, it is correct to spell the informal or colloquial word, "Neandertal."

A Note on *Homo sapiens afer*

H. sapiens afer is a name originally proposed by Linnaeus for black African people. Although that usage has long disappeared, the name was revived in the third quarter of the 20th century by Wells (1969, 1972) to refer to a subspecies in which to accommodate southern African human remains of Late Pleistocene age. This proposed subspecies was suggested to embrace populations deemed to be the common ancestor of the Khoisan and black (Negro) populations of present-day sub-Saharan Africa.

A number of human remains were assigned to this ancestral African population, such as those from Border Cave, Natal; Tuinplaas, Springbok Flats, of the Transvaal; and from Fish Hoek, Matjes River, and Klasies River Mouth Cave, of the southern Cape coast. The cranium from Florisbad, near Bloemfontein, Orange Free State, might have belonged to the same subspecies, although it shows some morphological reminders of *H. sapiens rhodesiensis*. Other remains considered to belong to the *afer* subspecies are known from Zambia, Kenya, and southern Ethiopia (Howell, 1978).

Morphologically, these *Homo sapiens afer* fossils show **gracilization** (that is, they are of slighter, more graceful body build than earlier *Homo*), and, in comparison with earlier hominids, their faces are reduced in size. Essentially, they are modern in form.

■

THE RISE OF ANATOMICALLY MODERN MAN

Although one of the oldest problems to excite the curiosity of natural historians, the subject of modern human origins underwent a resurgence of interest in the 1980s. This fresh spurt of attention was triggered by the use of new dating techniques, such as thermoluminescence, amino acid dating, electron spin resonance (ESR) dating, and uranium disequilibrium dating; by results of new molecular studies, and especially some startling claims from work on mitochondrial DNA; and by new comparative studies on Middle and Late Pleistocene hominid fossils. Today, in the 1990s, the subject of modern human origins has become one of the most exciting areas in paleoanthropology. One gauge of this renaissance is the fact that no fewer than four books have been devoted to the subject since 1984: These treatises have been edited by Smith and Spencer (1984), Mellars and Stringer (1989), Trinkaus (1989), and by Brauer and Smith (1990). In addition, a searching, dispassionate, and elegant appraisal of three "models" of modern human origins that have arisen since 1980 was published by Smith et al. (1989) in the *Yearbook of Physical Anthropology*. The three models have been designated by Smith and his colleagues as Brauer's Afro-European *sapiens* hypothesis, Stringer's and Andrew's recent African evolution model, and the multiregional evolutionary model of Wolpoff, Wu, and Thorne.

The subject of modern human origins is simply too vast to review in detail here. The main impetus to the new flurry of interest was provided by claims that several anatomically modern African hominids, most notably the fossils from Border Cave and from Klasies River Mouth Main Cave, in South Africa, were over 100,000 years old. Another such apparently early form of anatomically modern *H. sapiens* is known from the Kibish Formation of the Omo Basin in southwestern Ethiopia. Although some conflicting interpretations have been offered on its morphological affinities, the Kibish I hominid seems to resemble later Pleistocene representatives of *H. sapiens* (Howell, 1978; Rightmire, 1984). It is possible that these remains might be accommodated within the range of variation of *H. sapiens afer,* or they might represent another, as yet unnamed subspecies of *H. sapiens.* For these Kibish remains, too, an age of over 100,000 years has been proposed.

There was at first no evidence from elsewhere that anatomically modern man had appeared earlier than about 40,000 years B.P. If the African claims just noted are valid, they would seem to indicate that Africa contained the world's oldest known remains of anatomically modern *H. sapiens*. Of the anatomical modernity of this group of African specimens, there is very little doubt; but serious reservations have been expressed on the validity of the dating in most of the cases noted. Smith et al. (1989, p. 46), after a review of the evidence and the claims made, have concluded cautiously, ". . . even though dates toward the older portion of the range may be *conclusively* demonstrated in the future, the best that can presently be said is that modern *H. sapiens* appeared in southern Africa sometime between 40 **ky*** and 100 ky B.P.".

Recently, application of thermoluminescence and ESR dating techniques to the anatomically modern hominids of Qafzeh and Skhul, in northern Israel, has astonished paleoanthropologists by yielding dates of 115,000 to 81,000 years (Vandermeersch, 1981, 1989). The case for a date of about 90,000 years for the Qafzeh fossils (Bar-Yosef, 1990) is described by Smith et al. (1989, p. 48) as "the strongest for an antiquity of this magnitude for early modern *H. sapiens*."

It was at this stage of fossil hominid research that Cann and her colleagues (Cann, 1987, 1988; Cann et al., 1987), on mitochondrial DNA, and Wainscoat and his confrères (Hill and Wainscoat, 1986; Wainscoat et al., 1986), on nuclear DNA, offered genetic evidence for a relatively recent African origin of all modern human populations. Cann et al. attempted to calculate the rate of divergence reflected in various types of DNA occurring in **mitochondria** (the seat of aerobic respiration in eukaryotic cells; see Chapter 2) and, from this rate, they estimated that the ancestral stock of all European and Asian human populations diverged from the African stock between 180,000 and 90,000 years ago. This view suggests that "a wave of early modern humans arising in Africa reached and swept over Eurasia where they either outbred, outcompeted, outlasted, or exterminated the populations that they found there" (Simons, 1989; Stringer and Andrews, 1988). Because mitochondrial DNA is passed on through the mother's ovum, predominantly although not exclusively,** this theory became popularized as the "Eve hypothesis," and Africa became regarded as the new-old Eden! Those paleoanthropologists who support the idea that the anatomically modern *H. sapiens* erupted from Africa and, totally or in large measure, *replaced* existing hominid populations in all parts of the Old World, see support for their views in this genetic evidence.

A number of investigators have questioned the reasoning and the methodology of the genetic claims. It has been suggested by Trinkaus (cited by Simons, 1989) that if the calibration of the mitochondrial DNA clock is in error by a significant factor, then perhaps this procedure is documenting simply the spread of *H. erectus* out of Africa. This would have occurred at a time when there were, according to present views, no other hominids to be replaced anywhere else. Spuhler (1988) has studied the evolution of mitochondrial DNA in monkeys, apes, and humans, and has thrown serious doubt on the methodology used by Cann et al. For one thing, Spuhler points out that the "African" sample of Cann et al. (1987), which is crucial to their entire analysis, was limited to 20 individuals, of whom 18 were American-born blacks!

Moreover, the Cann et al. study of mitochondrial DNA in human populations was based on only about 1500 base pairs (the principal chemical components of DNA) out of some 16,500 such pairs that actually occur, less than 10% of the total. Spuhler (1988) believes that the mitochondrial DNA evidence gives more support to an hypothesis that accounts for the origin of "the major races of *H. sapiens*" by transitions occurring in three or more geographic regions, than to the hypothesis of worldwide replacement of prior stocks by migration from a single region.

---

*ky = one kiloyear (i.e., 1000 years).

**It is often overlooked that the spermatozoon bears mitochondria in the form of a helical sheath in its mid-piece. Because the mid-piece may enter an ovum during fertilization, this provides a mechanism for some "leakage" and for the addition of *paternal* mitochondrial DNA to that from the *maternal* source into the cytoplasm of the zygote.

Hence, at present, there are questions regarding both some of the critical geochronological, as well as the molecular biological interpretations. We clearly need more data on both of these aspects of the problem. It is a pity that so dramatic, so challenging, and so unexpected an hypothesis as that tracing modern humans to an African origin has encountered heavy weather—no other concept in recent decades has done more to enliven a previously neglected, and seemingly dull, area of the study of fossil mankind.

It was Thomas Henry Huxley, the ardent supporter of Charles Darwin, who wrote: "The great tragedy of science is the slaying of a beautiful hypothesis by an ugly fact." Is the reaction to the "Out of Africa" hypothesis an example of such a slaying? No, that would be going too far, for the fossil facts stand unassailed; not the fossils themselves, but only the dates for Border Cave, Klasies River Mouth, and for some of the others need to be confirmed. Whatever the final verdict on the genetic evidence, at the paleontological level it remains rather more likely that anatomically modern man emerged first in Africa, early in the Late Pleistocene, or even at the end of the Middle Pleistocene, than anywhere else. We may sum up with the modest and judicious interim conclusions of Smith et al. (1989):

1. A model that recognizes an important role for *local continuity* is the best explanation for modern human origins.

2. Significant genetic change was probably involved in the emergence of modern human anatomical form.

3. It is more logical to view this change as having occurred initially in one region, and then as having spread throughout the Old World.

4. This spread did not ubiquitously result from population migration.

5. New genetic elements were *assimilated* into existing gene pools or, sometimes, perhaps old elements were assimilated into new gene pools.

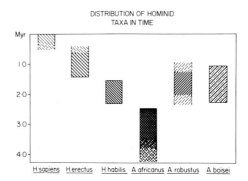

FIGURE 6.20

The approximate distribution in time of six major species of the Hominidae showing the temporal distribution of three species belonging to the genus *Australopithecus* and three species of the genus *Homo*. The two species on the right—*A. robustus,* from the Transvaal, South Africa; and *A. boisei,* from East Africa—are considered here to be on lineages which became extinct about 1 Myr ago. The two relatively early occurring species on the left—*H. habilis* and *H. erectus*—are regarded here as being on the lineage leading to the third species of *Homo,* namely, *H. sapiens. Australopithecus africanus* is the probable common ancestor both of the progressive forms (on the left) leading to modern man, and of the very specialized species (on the right) that led to an evolutionary cul-de-sac.

6. Biological reality is *not* best served by the notion that modern humans originated as the result of a speciation event.

7. There are still considerable gaps in our knowledge and a number of fundamental questions remain to be answered.

8. Each of the existing three models for the evolutionary emergence of modern humans has strengths and weaknesses, but none unequivocally explains *all* the available data.

In summarizing the hominid evolutionary phenomena during the Pleistocene Epoch (1.8 to 0.01 Myr B.P.), we see that in the Early Pleistocene (1.8 to 0.7 Myr B.P.), different hominid species were still competing with one another (*H. habilis*, *H. erectus*, *A. robustus*, and *A. boisei*; Figure 6.20). The Middle Pleistocene (0.7 to 0.125 Myr B.P.) was a time when *H. erectus* reached its evolutionary zenith and, according to the view supported here, became transformed, through *erectus*-to-*sapiens* intermediates, into archaic *H. sapiens*, which diversified into several subspecies in various parts of the Old World. During the Late Pleistocene (0.125 to 0.01 Myr B.P.), the newly emerged, anatomically modern *H. sapiens* held sole sway.

With regard to problems of evolutionary adaptation, biological mechanisms dominated hominid survival and evolution from the emergence of the Hominidae until about 2.5 Myr B.P.; thereafter, from 2.5 to about 1.5 Myr B.P., to an ever-increasing extent, cultural and behavioral mechanisms of adaptation were added to the biological ones; subsequently, it is reasonable to infer that, with stone toolmaking diversity, control of fire, widespread cave dwelling, powerful migratory ventures, and a hunting economy, cultural and behavioral

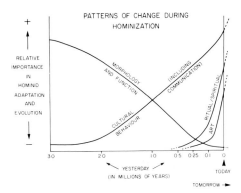

FIGURE 6.21

Diagrammatic summary of the changing pattern of biological and cultural traits, and their relative importance in hominid adaptation and evolution, during the most recent three million years of hominization. Morphological and functional factors were predominant determinants of survival at the time of *Australopithecus africanus*, about 3.0 Myr B.P. During subsequent hominization, such factors, although still manifestly significant, began to play a relatively less important role with the rise of cultural mechanisms of adaptation about 2.5 to 2.0 Myr B.P. Evidently, behavioral traits, including material culture and speech, gradually became overwhelmingly important in determining the adaptation and survival of the hominids, while anatomical and physiological adjustments played an ever-diminishing role. The diagram implies that in recent man, behavioral traits, including the most recent acquisitions (of ritual/spiritual and artistic dimensions), have come to play the most prominent role as determinants of man's present and future adaptation and development.

activities predominated in humankind's modes of survival and adaptation, from about 1.5 Myr B.P. up to the end of the Pleistocene and on into the Holocene (0.01 Myr B.P. until today; Figure 6.21).

Modern humankind has become *the* cultural primate par excellence—and what is more, they speak about it, and have even learned, quite recently, to write about it.

Acknowledgments

My sincere appreciation is extended to Dr. J. William Schopf and the Center for the Study of Evolution and the Origin of Life, UCLA, for making it possible for me to participate in the inaugural Annual CSEOL Symposium and to Dr. Schopf for his sensitive and punctilious editing of the manuscript. Very warm thanks are due to Mrs. Heather White for her thoughtful and diligent preparation of the typescript. I am most grateful to Mrs. Val Strong, Mr. Peter Faugust, and Mr. Terry Borain.

■

REFERENCES CITED

Bar-Yosef, O. 1990. Middle Palaeolithic chronology and the transition to the Upper Palaeolithic in Southwest Asia. In G. Brauer and F. H. Smith (Eds.), *Continuity or Replacement: Controversies in Homo sapiens Evolution* (Rotterdam: Balkema), pp. 8–37.

Boaz, N. T. 1979. Hominid evolution in eastern Africa during the Pliocene and early Pleistocene. *Ann. Rev. Anthropol. 8:* 71–85.

Boaz, N. T., and Burckle, L. H. 1984. Paleoclimatic framework for African hominid evolution. In J. C. Vogel (Ed.), *Late Cainozoic Palaeoclimates of the Southern Hemisphere* (Balkema: Rotterdam), pp. 483–498.

Brain, C. K. 1981. The evolution of man in Africa: Was it a consequence of Cainozoic cooling? *Trans. Geol. Soc. S. Afr. 84:* 1–19.

Brain, C. K., Churcher, C. S., Clark, J. D. Grine, F. E., Shipman, P., Susman, R. L., Turner, A., and Watson, V. 1988. New evidence of early hominids, their culture and environment from the Swartkrans Cave. *S. Afr. J. Sci. 84:* 828–835.

Brauer, G., and Smith, F. H. (Eds.). 1990. *Continuity or Replacement: Controversies in Homo sapiens Evolution* (Rotterdam: Balkema), 232 pp.

Cann, R. L. 1987. In search of Eve. *The Sciences* (Sept./Oct): 30–37.

Cann, R. L. 1988. DNA and human origins. *Ann. Rev. Anthropol. 17:* 127–143.

Cann, R. L., Stoneking M., and Wilson, A. C. 1987. Mitochondrial DNA and human evolution. *Nature 325:* 31–36.

Clark, LeGros, W. E. 1955. *The Fossil Evidence for Human Evolution: An Introduction to the Study of Paleoanthropology* (Chicago: Univ. of Chicago Press), 181 pp.

Clark, LeGros, W. E. 1964. *The Fossil Evidence for Human Evolution: An Introduction to the Study of Paleoanthropology* (2nd ed.) (Chicago: Univ. of Chicago Press), 201 pp.

Conroy, G. C. 1990. *Primate Evolution* (New York: Norton), 492 pp.

Conroy, G. C., Vannier, M. W., and Tobias, P. V. 1990. Endocranial features of *Australopithecus africanus* revealed by 2- and 3-D computed tomography. *Science 247:* 838–841.

Coon, C. S. 1963. *The Origin of Races* (London: Jonathan Cape), 724 pp.

Corvinus, G. 1976. Prehistoric exploration at Hadar, Ethiopia. *Nature 261:* 571–572.

Corvinus, G., and Roche, H. 1976. La préhistoire dans la région d'Hadar (Bassin de l'Awash Afar, Ethiopie): Premiers résultats. *L'Anthropologie 80:* 315–324.

Eccles, J. C. 1989. *Evolution of the Brain: Creation of the Self* (London: Routledge), 282 pp.

Falk, D. 1983. Cerebral cortices of East African early hominids. *Science 222:* 1072–1074.

Flohn, H. 1984. Climate evolution in the Southern Hemisphere and the equatorial region during the Late Cenozoic. In J. C. Vogel (Ed.), *Late Cainozoic Palaeoclimates of the Southern Hemisphere* (Rotterdam: Balkema), pp. 5–20.

Foley, R., and Dunbar, R. 1989. Beyond the bones of contention. *New Scientist* (14 Oct.): 37–41.

Gray, J. E. 1825. Outline of an attempt at the disposition of the Mammalia into tribes and families with a list of the genera apparently appertaining to each tribe. *Annals of Philosophy/New Series 10:* 337–344.

Goodman, M. 1974. Biochemical evidence on hominid phylogeny. *Ann. Rev. Anthropol. 3:* 203–228.

Grine, F. E. 1986. Dental evidence for dietary differences in *Australopithecus* and *Paranthropus:* A quantitative analysis of permanent molar microwear. *J. Hum. Evol. 15:* 783–822.

Harris, J. W. K. 1983. Cultural beginnings: Plio-Pleistocene archaeological occurrences from the Afar, Ethiopia. *Afr. Arch. Rev 2:* 3–31.

Hill, A. V. S., and Wainscoat, J. S. 1986. The evolution of the a- and b-globin gene clusters in human populations. *Hum. Gen. 74:* 16–23.

Holloway, R. L. 1975. Early hominid endocasts: Volumes, morphology and significance for hominid evolution. In R. H. Tuttle (Ed.), *Primate Functional Morphology and Evolution* (The Hague: Mouton), pp. 393–416.

Holloway, R. L., and Prost, D. G. 1982. The relativity of relative brain measures and hominid mosaic evolution. In E. Armstrong and D. Falk (Eds.), *Primate Brain Evolution: Methods and Concepts* (New York: Plenum), pp. 57–76.

Howell, F. C. 1978. Hominidae. In V. J. Maglio and H. B. S. Cooke (Eds.), *Evolution of African Mammals* (Cambridge: Harvard Univ. Press), pp. 154–248.

Jerison, H. J. 1973. *Evolution of the Brain and Intelligence* (New York: Academic Press), 482 pp.

Klein, R. G. 1988. The causes of "robust" australopithecine extinction. In F. E. Grine (Ed.), *Evolutionary History of the "Robust" Australopithecines* (New York: Aldine de Gruyter), pp. 499–505.

Leakey, L. S. B., Tobias, P. V., and Napier, J. R. 1964. A new species of the genus *Homo* from Olduvai Gorge. *Nature 202:* 7–9.

Leakey, M. D. 1967. Preliminary survey of the cultural material from Beds I and II, Olduvai Gorge, Tanzania. In W. W. Bishop and J. D. Clark (Eds.), *Background to Evolution in Africa* (Chicago: Chicago Univ. Press), pp. 417–446.

Leakey, M. D. 1971. *Olduvai Gorge,* Vol. III. *Excavations in Beds I and II, 1960–1963* (Cambridge: Cambridge Univ. Press), 306 pp.

Linnaeus, C. 1758. *Systema Naturae, Animalia* (Stockholm).

Lowrie, R. H. 1937. *The History of Ethnological Theory* (New York: Farrar and Rinehart), 293 pp.

Marshack, A. 1976. Some implications of the paleolithic symbolic evidence for the origin of language. *Curr. Anthropol. 17:* 274–282.

Mayr, E. 1981. Biological classification: Toward a synthesis of opposing methodologies. *Science 214:* 510–516.

McHenry, H. M. 1982. The pattern of human evolution: Studies on bipedalism, mastication and encephalization. *Ann. Rev. Anthropol. 11:* 151–173.

Mellars P., and Stringer C. B. (Eds.). 1989. *The Origin and Dispersal of Modern Humans: Behavioural and Biological Perspectives* (Edinburgh: Edinburgh Univ. Press), 419 pp.

Napier, J., and Tobias, P. V. 1964. The case for *Homo habilis. The Times,* London, 5 June.

Partridge, T. C. 1985. The palaeoclimatic significance of Cainozoic terrestrial stratigraphic evidence from South Africa: A review. *S. Afr. J. Sci. 81:* 245–247.

Rightmire, G. P. 1984. *Homo sapiens* in sub-Saharan Africa. In F. H. Smith and F. Spencer (Eds.), *The Origins of Modern Humans: A World Survey of the Fossil Evidence* (New York: Alan R. Liss), pp. 295–325.

Roche, H., and Tiercelin, J. J. 1977. Découverte d'une industrie lithique ancienne in situ dans la formation d'Hadar, Afar central Ethiopie. *C. R. Acad. Sci. Paris D284:* 1871–1874.

Simons, E. L. 1985. African origin, characteristics and context of earliest higher primates. In P. V. Tobias (Ed.), *Hominid Evolution: Past, Present and Future* (New York: Alan R. Liss), pp. 101–106.

Simons, E. L. 1989. Human origins. *Science 245:* 1343–1350.

Simpson, G. G. 1945. The principles of classification and a classification of mammals. *Bull. Amer. Mus. Nat. Hist. 85:* 1–350.

Skelton, R. R., McHenry, H. M., and Drawhorn, G. M. 1986. Phylogenetic analysis of early hominids. *Curr. Anthropol. 27:* 21–43.

Smith, F. H., Falsetti, A. B., and Donnelly, S. M. 1989. Modern human origins. *Yrb. Phys. Anthropol. 32:* 35–68.

Smith, F. H., and Spencer, F. (Eds.). 1984. *The Origins of Modern Humans: A World Survey of the Fossil Evidence* (New York: Alan R. Liss), 590 pp.

Spuhler, J. N. 1988. Evolution of mitochondrial DNA in monkeys, apes, and humans. *Yrb. Phys. Anthropol. 31:* 15–48.

Stringer, C. B. 1974. Population relationships of later Pleistocene hominids: A multivariate study of available crania. *J. Archaeol. Sci. 1:* 317–342.

Stringer, C. B., and Andrews, P. 1988. Genetic and fossil evidence for the origin of modern humans. *Science 239:* 1263–1268.

Taieb, M., and Tiercelin, J. J. 1979. Sedimentation Pliocene et paléoenvironments de rift: Exemple de la formation à hominides d'Hadar (Afar, Ethiopie). *Bull. Soc. Geol. France 21:* 243–53.

Tobias, P. V. 1964. The Olduvai Bed I hominine with special reference to its cranial capacity. *Nature 202:* 3–4.

Tobias, P. V. 1971. *The Brain in Hominid Evolution* (New York: Columbia Univ. Press), 170 pp.

Tobias, P. V. 1980. L'évolution du cerveau humain. *La Recherche 11:* 282–292.

Tobias, P. V. 1981. *The Evolution of the Human Brain, Intellect and Spirit* (Adelaide: Univ. of Adelaide), 70 pp.

Tobias, P. V. 1983. Recent advances in the evolution of the hominids with especial reference to brain and speech. *Pontif. Acad. Scient. Scripta Varia 50:* 85–140.

Tobias, P. V. 1984. Climatic change and evolution—evidence from the African fossil and hominid sites. In J. C. Vogel (Ed.), *Late Cainozoic Palaeoclimates of the Southern Hemisphere* (Rotterdam: Balkema), pp. 515–520.

Tobias, P. V. 1985. Ten climacteric events in hominid evolution. *S. Afr. J. Sci. 81:* 271–271.

Tobias, P. V. 1986. The last million years in Southern Africa. In T. Cameron and S. N. Spies (Eds.), *An Illustrated History of South Africa* (Johannesburg: Jonathan Ball), pp. 20–27.

Tobias, P. V. 1987. The brain of *Homo habilis:* A new level of organization in cerebral evolution. *J. Hum. Evol. 16:* 741–761.

Tobias, P. V. 1989. The environmental background of hominid emergence and the appearance of the genus *Homo. Int. J. Anthropol.* (in press).

Tobias, P. V. 1990a. *Olduvai Gorge,* Vols. 4A and 4B. *The Skulls, Endocasts and Teeth of Homo habilis* (Cambridge: Cambridge Univ. Press), 213 pp.

Tobias, P. V. 1990b. Man, culture and environment. In Proceedings of Symposium on "Evolution of Life," Kyoto, Japan (Tokyo: Springer-Verlag) (in press).

Trinkaus, E. (Ed.). 1989. *The Emergence of Modern Humans: Biocultural Adaptations in the Later Pleistocene* (Cambridge: Cambridge Univ. Press), 476 pp.

Vandermeersch, B. 1981. Les premiers *Homo sapiens* au Proche-Orient. In D. Ferembach (Ed.), *Les Processus de l'Hominisation* (Paris: CNRS), pp. 97–100.

Vandermeersch, B. 1989. The evolution of modern humans: Recent evidence from South-West Asia. In P. Mellars and C. B. Stringer (Eds.), *The Origins and Dispersal of Modern Humans* (Edinburgh: Edinburgh Univ. Press), pp. 21–47.

Vrba, E. S. 1985a. Early hominids in southern Africa: Updated observations on chronological and ecological background. In P. V. Tobias (Ed.), *Hominid Evolution: Past, Present and Future* (New York: Alan R. Liss), pp. 195–200.

Vrba, E. S. 1985b. Ecological and adaptive changes associated with early hominid evolution. In E. Delson (Ed.), *Ancestors: The Hard Evidence* (New York: Alan R. Liss), pp. 63–71.

Walker, A. 1981. Diet and teeth: Dietary hypotheses and human evolution. *Phil. Trans. R. Soc. London, B292:* 57–64.

Wallace, J. A. 1972. The dentition of the South African early hominids: A study of form and function. Ph.D. diss., Univ. of Witwatersrand (Dept. of Anatomy), Johannesburg, S. Africa.

Wallace, J. A. 1978. Evolutionary trends in the early hominid dentition. In C. J. Jolly (Ed.), *Early Hominids of Africa* (London: Duckworth), pp. 285–310.

Washburn, S. L., and Benedict, B. 1979. Non-human primate culture. *Man 14:* 163–164.

Wells, L. H. 1969. *Homo sapiens afer* Linn.—content and earliest representatives. *S. Afr. Archaeol. Bull. 24:* 172–173.

Wells, L. H. 1972. Late stone age and middle stone age tool-makers. *S. Afr. Archaeol. Bull. 27:* 5–9.

GLOSSARY

Acanthodia. Zoological group that includes early evolving spiny-finned fish.

Acanthomorph. Any of various types of acritarchs having surficial spiny processes (see *Acritarcha*).

Acritarcha. Artificial taxonomic group that includes Precambrian and Phanerozoic organic-walled, commonly spheroidal, "micro-algalike" microfossils of uncertain biologic relationships.

Actinopterygia. Zoological group that includes higher bony fish.

Adenine. Organic compound, a type of purine and one of the five nitrogen-containing bases present in biological nucleic acids (see *Nucleic acid*).

Adenosine triphosphate (ATP). Organic compound, in biological systems involved in the transfer of phosphate bond energy.

Adiabatic. Pertaining to the relationship of pressure and volume when a gas or fluid is compressed or expanded without either giving or receiving heat.

Aerobe. Organism capable of using molecular oxygen in aerobic respiration, either as a requirement (for "obligate aerobes") or facultatively (for "facultative aerobes"); an "aerobic environment" contains sufficient molecular oxygen to support growth of aerobic organisms (see *Anaerobe*).

Aerobic respiration. Molecular oxygen-using, energy-producing metabolic process carried out by obligate and facultative aerobes (see Table 2.1).

Aeronian Stage. Geological stage of the Lower Silurian.

Agnatha. Zoological group that includes living and extinct jawless fish.

Alete. Pertaining to spores that lack surficial sutures such as trilete marks (see *Trilete mark*).

Alga. Eukaryotic, photosynthetic, multicellular, nonvascular plant; "seaweeds" and their freshwater equivalents (see *Micro-alga* and *Protist*).

Alternation of generations. In sexually reproducing eukaryotes, pertaining to the alternation of diploid and haploid phases of the life cycle (see *Diploid* and *Haploid*).

Amino acid. Any of various organic compounds characterized by the presence of an amino group ($-NH_2$) and a carboxylic acid group ($-COOH$) of the type making up biological proteins.

Amniote egg. Yolk-bearing, usually hard-shelled egg typical of reptiles and birds; "amniotes" include mammals as well as reptiles and birds, but do not include fish or amphibians that possess "naked," yolkless eggs.

Amphibian. Tetrapod vertebrates such as toads, frogs, and salamanders that as adults are capable of living on land but that require an aqueous environment for reproduction.

Anaerobe. Organism that does not require molecular oxygen for metabolism, for example, an organism that lives solely via fermentation; an "anaerobic environment" does not contain sufficient molecular oxygen to support aerobic respiration (see *Aerobe*).

Anapsid. Any of various amniote tetrapods, such as turtles, characterized by a lack of temporal openings (fenestrae) in the skull (see *Diapsid* and *Synapsida*).

Angiospermae. Botanical group that includes flowering plants.

Angstrom (Å). Unit of length equal to one ten-billionth (1×10^{-10}) of a meter.

Anhydrous. Pertaining to the absence of water.

Ankylosauria. Zoological group that includes ornithischian armored dinosaurs (e.g., *Ankylosaurus*).

Anomodontia. Zoological group of herbivorous therapsid reptiles that includes the Eotitanosuchia, Dinocephalia, and Dicynodontia (see *Theriodontia*).

Anoxic. Pertaining to the absence of molecular oxygen.

Anoxic photosynthesis. Type of photosynthesis carried out by photosynthetic bacteria in which hydrogen from a source other than water is used to reduce carbon dioxide and via which molecular oxygen is therefore not released as a by-product (see *Oxygenic photosynthesis* and Table 2.1).

Antheridium. Specialized tissue of a gametophyte in which male gametes (sperm) are produced (see *Archegonium*).

Archean Era. Earliest era of the Precambrian Eon, extending from the formation of the earth to the beginning of the Proterozoic Era (see Geologic Timetable on inner front cover).

Archegonium. Specialized tissue of a gametophyte in which female gametes (eggs) are produced (see *Antheridium*).

Archetype. Hypothetical ancestral pattern or model of a particular group of organisms obtained by combining the primitive characteristics of living members of the group.

Archosauria. Major zoological group that includes such diapsid reptiles as crocodiles, alligators, dinosaurs, pterosaurs, and thecodonts.

Asexual. Pertaining to a life cycle that lacks sexual reproduction.

Asteroid. Any of the small "planetoid" bodies in orbit about the sun.

Autocatalysis. Modification, especially an increase in rate, of a chemical reaction by the product of that reaction.

Autotroph. "Plantlike" (including prokaryotic) organism capable of using carbon dioxide as its sole carbon source (see *Heterotroph* and Table 2.1).

Axis. General term pertaining to commonly cylindrical, "stemlike" (erect) and/or "rootlike" (prostrate), water- and photosynthate-conducting plant organs.

Bacterium. Any of diverse types of prokaryotes (including Eubacteria, Archaebacteria, and prochlorophytes) other than cyanobacteria.

Banded iron-formation (BIF). Chemically deposited cherty sedimentary rock, typically thinly bedded and containing more than 15% iron.

Bilateria. Major zoological group that includes all metazoans except sponges and cnidarians.

Biostratigraphic zone. Subdivision of a geologic unit characterized by the presence of a specific fossil taxon or suite of taxa.

Bisexual. Pertaining to a dyad or tetrad the spores of which germinate to produce both male and female gametophytic plants.

Blue-green algae. Informal name for cyanobacteria.

Bradytely. Unusually slow rate of morphological evolution (see *Horotely, Hypobradytely*, and *Tachytely*).

Bryophyte. Nonvascular land plant such as a liverwort, hornwort, or moss.

Butlerow synthesis. Chemical reaction sequence resulting in synthesis of sugars from formaldehyde, known also as the "formose reaction."

Calcareous. Containing or composed of calcium carbonate.

Caliche. Carbonate cemented calcareous soil crust developed typically in arid or semiarid regions.

Cambrian Period. Earliest geological period of the Phanerozoic Eon (see Geologic Timetable on the inner front cover).

Captorhinid. Any of numerous lizardlike, early evolving, anapsid tetrapods.

Carbonaceous. Containing or composed of kerogenous organic matter, whether of abiological (e.g., organic matter in carbonaceous chondrites) or biological origin (e.g., coal).

Carbonaceous chondrite. Stony meteorite characterized by the presence of chondrules, spheroidal glassy granules usually about 1 mm in diameter that contain carbonaceous matter.

Carbonate. Mineral containing the carbonate chemical group, CO_3 (e.g., calcite, $CaCO_3$), or a rock consisting chiefly of carbonate minerals (e.g., limestone and dolostone).

Carboniferous Period. Late Paleozoic geological period, used particularly in Europe, equivalent to the combined Mississippian and Pennsylvanian Periods of North America (see Geologic Timetable on inner front cover).

Cardoc Stage. Geological stage of the Upper Ordovician.

Carnivore. Predominantly meat-eating animal (see *Herbivore* and *Omnivore*).

Cartilage. Elastic noncellular tissue that composes the skeleton of embryonic and very young vertebrates and becomes for the most part converted into bone in adult higher vertebrates.

Cast. Secondary sedimentary material that infills a cavity (natural mold) resulting from decay or dissolution of part or all of an organism, replicating details of its external form (see *Mold*).

Catalysis. Modification, especially an increase of rate, of a chemical reaction induced by material that is unchanged chemically at the end of the reaction.

Cenozoic Era. Latest of three eras of the Phanerozoic Eon, extending from the end of the Mesozoic Era to the present (see Geologic Timetable on the inner front cover).

Cephalaspid. Any of various agnathid fish characterized by paired pectoral finlike appendages and bony head/thoracic armor.

Ceratopsia. Zoological group that includes horned dinosaurs (e.g., *Triceratops*).

Charophyceae. Botanical family of green algae, informally referred to as "stoneworts."

Chert. Chemically deposited sedimentary rock consisting chiefly of very fine-grained quartz.

Chloroplast. Chlorophyll-containing, membrane-enclosed organelle of plants and photosynthetic protists; the site of photosynthesis in eukaryotic cells.

Chordata. Major zoological group that includes animals having at some stage in their development a notochord and a dorsal nervous system, with or without a segmented backbone.

Chromatography. Process in which a chemical mixture carried by a liquid or gas is separated into its identifiable chemical components.

Chronospecies. Fossil species defined by their temporal distribution.

Clade. Group of taxa derived from a single common ancestor.

Cladistics. Type of biological systematics based on character analysis.

Cladogenic. Event resulting in phylogenetic branching and development of a new clade.

Cladogram. Branching diagrammatic tree used in cladistics to illustrate phylogenetic relationships.

Class. In biological classification, a major category ranking above the order and below the (zoological) phylum or the (botanical) division.

Club moss. Informal name for lycopods, members of the Lycophytina (e.g., *Lycopodium*, *Selaginella*).

Cnidaria. Zoological group that includes hydrozoans, jellyfish, sea anemones, and corals; known also as the Coelenterata.

Coelacanth. Any of a zoological family of largely extinct lobe-finned fish (e.g., *Latimeria*).

Coelomate. Animal that possesses a coelom, the usually epithelium-lined space between the body wall and the digestive tract in higher metazoans.

Cold-blooded. Informal term for ectothermy.

Collagen. Fibrous, hydroxy proline-rich protein, the connective tissue characteristic of metazoans.

Compression fossil. Flattened carbonaceous fossils, preserved typically in shales and claystones.

Conglomerate. Coarse-grained sedimentary rock formed of rounded fragments less than 2 mm in diameter and cemented in a finer grained matrix; consolidated equivalent of gravel.

Consumer. Any of various heterotrophs of an ecosystem.

Cortex. In plant axes, the tissue occurring between the external surface (epidermis) and the vascular tissue.

Cretaceous Period. Youngest geological period of the Mesozoic Era (see Geological Timetable on the inner front cover).

Cryptogam. Any of diverse types of spore-producing plants.

Cryptospore. Plant spore, occurring in more or less permanent tetrads, or in dyads or as monads derived from such tetrads, particularly characteristic of Ordovidian and Lower Silurian strata.

Curvaturae. The arcuate margin of contact of a spore in a tetrahedral tetrad with the other three spores of the tetrad.

Cuticle. Waxy (cutinized) layer formed on the outer (epidermal) wall of axes, leaves, and similar plant organs.

Cynodontia. One of three zoological subgroups of theriodont therapsid reptiles.

Cyst. Microscopic saclike body with a resistant wall that in some micro-algae and other protists serves as a resting spore preceding development of a specialized reproductive phase.

Cytosine. Organic compound, a type of pyrimidine and one of the five nitrogen-containing bases present in biological nucleic acids (see *Nucleic acid*).

Dehydration condensation. Chemical reaction involving formation of a water molecule (HOH) by removal of hydrogen (−H) and hydroxyl (−OH) from two monomers that become combined into a dimer, or from one monomer and an oligomer that become combined into a polymer.

Dendrogram. Branching diagrammatic tree used to illustrate evolutionary relationships.

Deoxyribonucleic acid (DNA). Usually helically wound, double-stranded nucleic acid, the genetic material of prokaryotes and eukaryotes, composed of four types of nitrogen-containing bases (adenine, guanine, thymine, and cytosine) bound to a linear sugar- (deoxyribose-) phosphate "backbone."

Deoxyribose. The five-carbon sugar occurring in DNA.

Deuterostomia. Major zoological group which includes animals such as echinoderms and chordates that share a particular type of developmental pathway.

Devonian Period. Geological period of the Paleozoic Era (see Geological Timetable on inner front cover).

Diapsid. Any of various reptiles, such as crocodiles, characterized by the presence of two pairs of temporal openings (fenestrae) in the skull (see *Anapsid* and *Synapsida*).

Dicarboxylic acid. Type of organic compound containing two carboxylic acid (−COOH) groups.

Dichotomous. The bifurcating (V-shaped) branching pattern characteristic of the axes of very early evolving vascular plants.

Dicynodontia. One of three zoological subgroups of anomodont therapsid reptiles that includes, for example, *Lystosaurus*.

Dinocephalia. One of three zoological subgroups of anomodont therapsid reptiles that includes, for example, *Moschops*.

Dioecious. Having the male and female organs on two separate plants (see *Monoecious*).

Diploid. Referring to the double chromosome number ("2N") in sexual eukaryotes, characteristic of the life cycle generation derived by mitotic division of a zygote (e.g., the sporophyte generation in plants and the embryo to adult in animals; see *Haploid*).

Dispersed spores. Spores that have been dispersed from a sporangium (see *In situ spores*).

Distal face. Of spores in a tetrahedral tetrad, the side of the spore on the exterior of the tetrad (see *Proximal face*).

Dominant generation. In sexually reproducing eukaryotes, the more prominent of the two life cycle generations (e.g., in animals and vascular plants, the diploid generation derived by mitotic division of a zygote; see *Alternation of generations* and *Diploid*).

Dyad. Paired spores, produced by partial fragmentation of a tetrad.

Ectoderm. The outer cellular membrane of two cell-layered animals (e.g., cnidarians) or the outermost of the three embryonic cell layers of higher animals (see *Endoderm* and *Mesoderm*).

Ectothermy. Behaviorally controlled thermoregulation involving uptake of heat from the environment (see *Endothermy*).

Ediacara fauna. "Soft-bodied" (biologically unmineralized) biotic assemblage characteristic of the Vendian Period of the Proterozoic Era.

Eifelian Stage. Geological stage of the Middle Devonian.

Embryophyte. Any of diverse higher land plants, including bryophytes and tracheophytes.

Emsian Stage. Geological stage of the Lower Devonian.

Enations. Bumplike or spiny epidermal protrusions present on some trimerophyte and zosterophyllophyte axes, regarded as evolutionary precursors of lycopod microphylls.

Encalyptaceae. Botanical family of the mosses.

Endocranial cast. Cast of a skull cavity, commonly providing evidence of the surficial morphological details of the originally enclosed soft tissue (see *Cast* and *Mold*).

Endoderm. The inner cellular layer of two cell-layered animals (e.g., cnidarians) or the innermost of the three embryonic cell layers of higher animals (see *Ectoderm* and *Mesoderm*).

Endolith. Organism (commonly prokaryotic, micro-algal, or fungal) living within consolidated rock or a soil crust.

Endophyte. Photosynthetic organism (commonly cyanobacterial or micro-algal) living within the tissue of a host plant.

Endosymbiosis. Symbiosis in which a symbiont lives within the body of its symbiotic partner.

Endothermy. Type of thermoregulation based on internally generated metabolic heat production (see *Ectothermy*).

Enzyme. Any of numerous complex proteins that catalyze specific biochemical reactions.

Eotitanosuchia. One of three zoological subgroups of anomodont therapsid reptiles.

Epiphyte. Plant which grows upon another plant, gaining physical support from it.

Equilibrium mixture. Chemical mixture in which the component phases do not undergo any net change in properties with the passage of time.

Ethology. The scientific study of animal behavior, especially under natural conditions.

Eukaryote. Unicellular or multicellular organism (viz., a protist, fungus, plant, or animal) characterized by nucleus-, mitochondrion-, and (in plants and photoautotrophic protists) chloroplast-containing cells that are capable typically of mitotic cell division (see *Prokaryote* and Table 2.2).

Eumetazoa. Major zoological group that includes all multicellular animals except the sponges.

Eurypterid. Any of the aquatic, usually large, Paleozoic arthropods included in the extinct zoological group Eurypterida.

Exine. The outer coat or wall of a higher plant spore.

Exoexine. The outer smooth or sculptured layer of a spore exine (see *Intexine*).

Facultative aerobe. Prokaryote capable of aerobic respiration but also able to grow in the absence of molecular oxygen via anaerobic processes (see *Obligate aerobe*).

Facultative biped. Tetrapod capable of walking either on two or four feet.

Family. In biological classification, a major category ranking above the genus and below the order.

Fecal pellet. Organic excrement, mainly of invertebrates, usually of simple ovoid form.

Fenestrae. Openings in body tissue, such as those present in temporal regions of diapsid or synapsid skulls.

Fermentation. Energy-producing metabolic process carried out in the absence of molecular oxygen (see Table 2.1).

Fertile axis. Plant axis on which one or more sporangia are borne (see *Sterile axis*).

Filament. In microbiology, a collective term referring to the cylindrical external sheath and the cellular internal trichome of a filamentous prokaryote.

Flavonoid. Any of a group of multiple-ring organic compounds that includes many commonly occurring pigments.

Formaldehyde. Organic compound, the simplest aldehyde, HCHO, an important intermediate compound in various prebiotic syntheses.

Formation. The fundamental formal unit in the stratigraphic classification of rocks, a characterizable mappable body of (igneous, sedimentary, or metamorphic) rock.

Frasnian Stage. Geological stage of the Middle Devonian.

Free-sporing. Pertaining to the release of spores from the sporangium as opposed to their retention in strobili ("cones") of some spore-producing plants.

Frond. Relatively large leaf (as in palms or ferns) or flat structure of leaflike form.

Gamete. Haploid female (egg) and male (sperm) sex cells, in animals produced via meiotic division and in plants via mitotic division in the gametophyte generation (see *Alternation of generations* and *Haploid*).

Gametophyte. In a plant life cycle, the haploid generation in which gametes are produced (see *Alternation of generations* and *Sporophyte*).

Gedinnian Stage. Geological stage of the Lower Devonian.

Gene. That segment of the genetic material (DNA) coding for a single protein molecule (the "one gene = one enzyme" concept of Beadle and Tatum).

Genome. The total genetic material of an organism.

Genotype. The total genetic information of an organism (see *Phenotype*).

Genus. In biological classification, a major category ranking above the species and below the family.

Givetian Stage. Geological stage of the Middle Devonian.

Glucose. Organic compound, the six-carbon sugar $C_6H_{12}O_6$ (abbreviated "CH_2O").

Glycoside. Any of numerous sugar derivatives that contain a nonsugar chemical group and that on hydrolysis yield a sugar.

Gondwanaland. Late Paleozoic to early Mesozoic supercontinent composed of the land masses of South America, Africa, Madagascar, India, and Antarctica.

Gorgonopsia. One of three zoological subgroups of theriodont therapsid reptiles that includes, for example, *Lycaenops*.

Gracile. With reference to a relatively small, graceful body.

Greenhouse effect. Warming of the earth's surface and the lower layers of the atmosphere caused by the interaction of solar radiation with atmospheric gases (principally carbon dioxide, methane, and water vapor) and its conversion into heat.

Group. Major category in the stratigraphic classification of rocks ranking above the formation and below the supergroup.

Guanine. Organic compound, a type of purine and one of the five nitrogen-containing bases present in biological nucleic acids (see *Nucleic acid*).

Gymnospermae. Botanical group that includes plants with naked seeds such as conifers, cycads, and *Gingko*.

Hadrosauridae. Zoological family of "duck-billed" dinosaurs.

Half-life. The time required for half of something to undergo a process, such as the time required for half of the atoms of an amount of radioactive isotope to become disintegrated.

Haploid. Referring to the single chromosome number ("1N") in sexual eukaryotes, characteristic of the life cycle generation derived from meiotically produced spores or produced directly by meiosis (e.g., the gametophyte generation in plants and the gametes in animals; see *Diploid*).

Hematite. Mineral, the iron oxide Fe_2O_3.

Hepatic. Informal name for liverworts.

Herbivore. Predominantly plant-eating animal (see *Carnivore* and *Omnivore*).

Heterotroph. "Animal-like" (including prokaryotic) organism capable of using organic matter as its sole carbon source (see *Autotroph* and Table 2.1).

Heterotrophic hypothesis. Concept introduced by A. I. Oparin and J. B. S. Bernal that the earliest forms of life were heterotrophs, using abiologically produced organic matter as their carbon source (also referred to as the "Oparin-Haldane hypothesis").

Holocene Epoch. Most recent geological epoch, also referred to as the Recent Epoch (see Geological Timetable on the inner front cover).

Homerian Stage. Geological stage of the Silurian.

Hominidae. Zoological family that includes humans; hominids.

Hominoidea. Zoological superfamily that includes humans, gibbons, and great apes; hominoids.

Homologous. Phenotypic characters that are of shared evolutionary derivation.

Homosporous. Cryptogammic plants that produce a single size of spore (see *Microspore* and *Megaspore*).

Horotely. Standard, usual rate of morphological evolution (see *Bradytely, Hypobradytely,* and *Tachytely*).

Hydrocarbon. Any of a diverse group of organic compounds composed of hydrogen and carbon.

Hydrogen cyanide. Organic compound, HCN.

Hydroid. In bryophytes, tubular water-conducting tissue.

Hydrolysis. Chemical process of decomposition involving addition of the elements of water (HOH).

Hydrophobic. Lacking affinity for water.

Hydroxy acid. Any of various acidic (typically COOH-containing) organic compounds characterized by the presence of the hydroxyl $(-OH)$ group.

Hypobradytely. Exceptionally slow rate of morphological evolution, characteristic of cyanobacteria (see *Bradytely, Horotely,* and *Tachytely*).

Ichthyosaur. Any of an extinct group (the Ichthyosauria) of marine fishlike or porpoiselike reptiles abundant in Mesozoic seas.

Imidazole. Organic compound, the heterocyclic base $C_3H_4N_2$.

Incertae sedis. Of uncertain systematic position.

In situ spores. Spores occurring within a sporgangium (see *Dispersed spores*).

Intermediate compound. Compound formed in a chemical reaction series prior to formation of the final product.

Intexine. Inner layer of a spore exine (see *Exoexine*).

Invertebrate. Animal lacking a backbone; "lower metazoan."

Isomer. One of two or more chemical compounds that contain the same numbers of atoms of the same elements but that differ in structural arrangement and properties.

Isospore. With reference to spores in a sporangium all being of similar size.

Isotope. Any of two or more types of atoms of a chemical element having nearly identical chemical behavior but with differing atomic mass and physical properties.

Jurassic Period. Mid-Mesozoic geological period (see Geological Timetable on the inner front cover).

Kerogen. Particulate, geochemically altered, macromolecular organic matter, insoluble in organic solvents and mineral acids, present in sedimentary rocks.

Ky. One kiloyear, 1000 years.

Labyrinthodont. Extinct tetrapods, the earliest known (Devonian and Carboniferous) amphibians, including the ichthyostegids, anthracosaurs, and temnospondyls.

Lagerstätte. Any of various sedimentary deposits unusual because of the exceptionally well-preserved fossils they contain, such as the Cambrian Burgess Shale and the Jurassic Solnhofen Limestone.

Landovery Stage. Geological stage of the Lower Silurian.

Land plant. Terriginous "higher plants," the bryophytes and tracheophytes.

Lateral sporangia. Spore sacs borne on the sides of plant axes, as in zosterophyllophytes and lycophytes (see *Terminal sporangia*).

Leptoid. Tubular photosynthate-conducting tissue in bryophytes.

Lichen. "Plantlike" symbiotic association of algae and fungi.

Lignin. Polyphenolic organic compound characteristically occurring in the xylem of vascular plants.

Ludfordian Stage. Geological stage of the Upper Silurian.

Ludlow Stage. Geological stage of the Upper Silurian.

Lycophytina. Major botanical group (subdivision) that includes clubmosses.

Ma. Mega anna, one million (1×10^6) years.

Magnetite. Mineral, the iron oxide Fe_3O_4.

Marchantiales. Botanical order of the liverworts.

Mass spectrometry. Instrumental method of identifying the chemical constituents of a substance by means of the separation of gaseous ions according to their differing mass and charge.

Megaphyll. "Fern-type" leaves, commonly with a network pattern of venation, characteristic of the Pterophytina (see *Microphyll*).

Megasphaeromorph. Spheroidal acritarchs larger than 200 μm in diameter.

Megaspore. Larger spores of plants producing two sizes of spores.

Meiosis. Type of division of the cell nucleus resulting in formation of four daughter cells (spores or their equivalents) each containing the haploid number of chromosomes; in eukaryotes, meiosis and syngamy (fusion of gametes) are fundamental aspects of sexual reproduction (see *Mitosis*).

Melting point. Temperature at which a substance becomes altered from a solid to a liquid state.

Mesoderm. One of three embryonic cell layers of higher animals, situated between the ectoderm and the endoderm.

Mesozoic Era. One of three eras of the Phanerozoic Eon, intermediate in age between the Paleozoic and Cenozoic Eras (see Geological Timetable on the inner front cover).

Messinian crisis. Late Miocene phase of cooling and increasingly arid climate that resulted in drying up of the Mediterranean Basin and the interchange of African and Asian faunas.

Metamorphic rock. Rock derived as a result of mineralogical, chemical, and structural changes in a preexisting rock in response to changes in temperature, pressure, and chemical environment at depth in the earth's crust.

Metazoan. Multicellular heterotrophic animal, commonly megascopic and mobile.

Micro-alga. Microscopic photoautotrophic protist, especially unicellular eukaryotic phytoplankton.

Micrometer (μm). Unit of length equal to one millionth (1×10^{-6}) of a meter.

Microphyll. Univeined leaves characteristic of the Lycophytina and Sphenophytina (see *Megaphyll*).

Microspore. Smaller spores of plants producing two sizes of spores.

Mid-Mesozoic evolutionary gap. The gap in the evolution of mammals between their origin and their rise to dominance.

Miocene Epoch. Epoch of the Tertiary Period (see Geological Timetable on the inner front cover).

Mitochondrion. Organelle of eukaryotic cells, the site of aerobic respiration.

Mitosis. Type of division of the cell nucleus resulting in formation of two daughter cells, each of which is an exact copy of the parent cell; in unicellular eukaryotes, a type of asexual reproduction (see *Meiosis*).

Mold. The cavity in sedimentary material resulting from decay or dissolution of part or all of an organism which upon infilling by secondary sediment results in formation of a cast (see *Cast*).

Monoecious. Having the male and female organs on the same plant (see *Dioecious*).

Monomer. Chemical compound that can undergo polymerization (see *Polymer*).

Mononucleotide. Nucleotide composed of one monomer each of a nitrogen-containing base, a sugar, and a phosphoric acid.

Monophyletic. Of or relating to a single evolutionary lineage, derived from a common ancestral stock (see *Polyphyletic*).

Monopodial Branching pattern in plants in which subsidiary branches emerge from a single major axis, as in the Lycophytina and Pterophytina.

Mosaic evolution. Pattern of evolution in which different structures or organs of the body evolve at different rates.

Mosasaur. Any of the zoological family Mosasauridae, extinct fish- or ammonite-eating aquatic reptiles of the lizard group.

Moss. Any of a botanical class (Musci) of bryophytic plants.

Myr B.P. Million (10^6) years before the present.

Naked axis. Leafless plant axis, such as those of rhyniophytoids and rhyniophytes.

Nitrogen fixation. The metabolic assimilation of atmospheric nitrogen by various prokaryotes.

Notochord. Longitudinal flexible rod of cells that in the lowest chordates (e.g., amphioxus) and in the embryos of the higher vertebrates forms the supporting axis of the body.

Nucleic acid. Any of various organic acids (e.g., DNA, RNA) formed of a sugar- (deoxyribose- or ribose-) phosphate "backbone" with attached purine (viz., adenine and/or guanine) and pyrimidine (viz., cytosine, thymine, and/or uracil) nitrogen-containing bases (see *Deoxyribonucleic acid* and *Ribonucleic acid*).

Nucleoside. Chemical compound that consists of a purine or a pyrimidine nitrogen-containing base combined with deoxyribose or ribose sugar (see *Nucleotide*).

Nucleotide. Chemical compound that consists of a purine or a pyrimidine nitrogen-containing base combined with deoxyribose or ribose sugar and with a phosphate group (see *Nucleoside*).

Obligate aerobe. Organism metabolically dependent on aerobic respiration and therefore on the availability of molecular oxygen (see *Facultative aerobe*).

Oligomer. Small polymeric chemical compound composed of a few or several monomers.

Omnivore. Any of various heterotrophs that eat both plant and animal material (see *Carnivore* and *Herbivore*).

Ontogeny. The developmental life cycle of an individual organism.

Oparin-Haldane hypothesis. See *Heterotrophic hypothesis*.

Order. In biological classification, a major category ranking above the family and below the class.

Ordovician Period. Geological period of the Paleozoic Era (see Geological Timetable on the inner front cover).

Organelle. Membrane-enclosed bodies, such as mitochondria and chloroplasts, that carry out specific biochemical functions and that occur in eukaryotic cells.

Organic compound. Chemical compound of the type typical of, but not restricted to, living systems, composed commonly of carbon, hydrogen, oxygen, nitrogen, sulfur, and/or phosphorus.

Ornithischia. Zoological group of dinosaurs characterized by a birdlike pelvis (see *Saurischia*).

Ornithopoda. Zoological group of facultative bipedal ornithischian dinosaurs such as, for example, *Hadrosaurus*.

Ossify. To change into or to form bone.

Ostracoderm. Any of various early evolving small agnathid fishes characterized by a lack of a lower jaw and usually a lack of paired fins and exhibiting a cartilaginous internal skeleton and, in some taxa, a bony head shield.

Oviparous. Producing eggs that mature outside the maternal body (see *Ovoviviparous* and *Viviparous*).

Ovoviviparous. Producing eggs that mature within the maternal body (see *Oviparous* and *Viviparous*).

Oxic. Pertaining to the presence of molecular oxygen.

Oxidize. To combine chemically with oxygen.

Oxygenic photosynthesis. Type of photosynthesis carried out by cyanobacteria and photoautotrophic eukaryotes in which hydrogen from water is used to reduce carbon dioxide and via which molecular oxygen is therefore released as a by-product (see *Anoxic photosynthesis* and Table 2.1).

Oxygen sink. Molecular oxygen-consuming process (such as aerobic respiration) or substance (such as unoxidized volcanic gases or dissolved ferrous iron).

Pachycephalosauria. Zoological group that includes ornithischian "dome-headed" dinosaurs such as *Stegoceras*.

Paleozoic Era. Earliest era of the Phanerozoic eon (see Geological Timetable on the inner front cover).

Palynology. The scientific study of plant pollen and spores, especially those of fossil plants.

Parsimony. In phylogenetics, economy of explanation of biological relationships.

Pelycosaur. Any of a zoological group (the Pelycosauria) that includes primitive synapsid "mammal-like" reptiles, for example, *Dimetrodon* (see *Synapsida*).

Pennsylvanian Period. Geological period of the Paleozoic Era recognized in North America and equivalent to the Late Carboniferous as recognized in Europe (see Geological Timetable on the inner front cover).

Peptide. Type of chemical bond that links adjacent amino acids in proteins, or a chemical compound, such as a protein, that contains such bonds.

Permian Period. Latest geological period of the Paleozoic Era (see Geological Timetable on the inner front cover).

pH. Scale from 0 to 14 used in chemistry to express both acidity and alkalinity in which a value of 7 represents neutrality, values less than 7 increasing acidity, and values greater than 7 increasing alkalinity.

Phanerozoic Eon. The younger of the two geological eons of earth history, extending from the beginning of the Cambrian Period to the present (see Geological Timetable on the inner front cover).

Phenotype. The total physical and biochemical traits of an organism, the physical and chemical expression of the genotype.

Phloem. Thin-walled, tubular, photosynthate-conducting tissue of a vascular plant (see *Vascular tissue* and *Xylem*).

Photic zone. Zone into which light energy penetrates that can be used by photosynthetic organisms.

Photoautotroph. Autotroph that uses energy from light, such as a photosynthetic bacterium, cyanobacterium, or plant.

Photochemical. Of, relating to, or resulting from the chemical action of light energy.

Photodissociation. Dissociation of a chemical compound under the influence of light energy, as in the dissociation of water (HOH) into hydrogen ($-$H) and hydroxyl ($-$OH).

Photolysis. Decomposition of a chemical compound under the influence of light energy.

Photosynthesis. See *Anoxic photosynthesis* and *Oxygenic photosynthesis*.

Photosynthetic bacterium. Any of diverse prokaryotes capable of anoxic photosynthesis.

Phototaxis. Movement of an organism in response to a gradient of light.

Phylogeny. The evolutionary group history of a kind of organism.

Phylum. In zoological classification, a major category ranking above the class; equivalent to a botanical division.

Plate tectonics. Global tectonics based on an earth model characterized by many thick oceanic or continental plates that move slowly across the global surface propelled by movement of underlying material of the planetary interior.

Pleisiomorphic. In cladistic analysis, a primitive character shared by members of a clade.

Pleistocene Epoch. Geological epoch, characterized by widespread continental glaciation, of the Quaternary Period (see Geological Timetable on the inner front cover).

Plesiosaur. Any of a zoological group (the Plesiosauria) of marine Mesozoic reptiles.

Pliocene Epoch. Geological epoch of the Tertiary Period (see Geological Timetable on the inner front cover).

Polymer. Chemical compound formed by polymerization and composed of repeating monomeric structural units (see *Monomer*).

Polymerase. Any of several enzymes that catalyze the polymerization of nucleic acids (DNA, RNA) from precursor substances.

Polynucleotide. Polymer composed of nucleotides.

Polyphenolic. Pertaining to a polymer containing, or wholly composed of, phenol (C_6H_5OH) subunits.

Polyphyletic. Of or relating to more than one evolutionary lineage, derived from two or more ancestral stocks (see *Monophyletic*).

Pongidae. Zoological family that includes the apes; pongids.

Precambrian Eon. The older of the two geological eons of earth history, extending from time of formation of the earth to the beginning of the Phanerozoic Eon (see Geological Timetable on the inner front cover).

Priapulida. Zoological phylum that includes priapulid worms.

Pridolian Stage. Geological stage of the Upper Silurian.

Problematica. Organism of uncertain biological relationships.

Progymnospermopsida. Botanical group that includes extinct spore-producing plants having fernlike foliage and gymnosperm- (conifer-) like wood.

Prokaryote. Unicellular or colonial microbial microorganism (viz., an archaebacterium, eubacterium, prochlorophyte, or cyanobacterium) characterized by cells that lack membrane-enclosed organelles such as nuclei, mitochondria, or chloroplasts and that reproduce via non-mitotic and nonmeiotic division (see *Eukaryote* and Table 2.2).

Protein. Polymeric organic compound composed of amino acids; polypeptide.

Proterozoic Era. The younger of two eras of the Precambrian Eon (see Geological Timetable on the inner front cover).

Protist. Any of diverse types of unicellular animal-like (e.g., protozoan) or plantlike (e.g., micro-algal) eukaryotes.

Protolignin. Informal term applied to polyphenolic "ligninlike" organic compounds occurring in some bryophytes.

Proximal face. Of spores in a tetrahedral tetrad, the side of the spore on the interior of the tetrad, adpressed against the other three spores (see *Distal face*).

Pseudofossil. Naturally occurring structure or object of nonbiological origin that bears resemblance to, and can be mistaken for, a true fossil.

Pseudosaccate spore. Plant spore characterized by its enclosure in a saclike envelope that is attached proximally to the inner body of the spore.

Pterosaur. Any of a zoological group (the Pterosauria) of Mesozoic flying reptiles.

Purine. Any of various nitrogen-containing bases, such as adenine and guanine, with the empirical formula $C_5H_4N_4$.

Pyrimidine. Any of various nitrogen-containing bases, such as cytosine, thymine, and uracil, with the empirical formula $C_4H_2N_2$.

Pyrite. Mineral, the iron sulfide FeS_2.

Pyrolysis. Chemical change, commonly decomposition, brought about by the action of heat.

Quartz. Mineral, crystalline silica, SiO_2.

Recent Epoch. Epoch of the Quarternary Period, equivalent to the Holocene Epoch (see Geological Timetable on the inner front cover).

Red bed. Sandstone or shaley sedimentary rock, commonly of terrigenous origin, having a reddish color imparted by iron oxides.

Reduce. In chemistry, to combine with or subject to the action of hydrogen, thereby lowering the state of oxidation.

Relative brain weight. Weight of the brain expressed as a fraction of the weight of the body.

Reptile. Type of amniote tetrapod, any of diverse members of a major zoological category (the Reptilia) that includes lizards, snakes, turtles, and crocodiles, as well as dinosaurs, pterosaurs, and related nonavian and nonmammalian vertebrates.

Retusoid spores. Plant spores characterized by a rounded or obtuse apex.

Rhyniophytina. Major botanical group (subdivision) that includes early evolving vascular plants (e.g., *Rhynia*) having dichotomously branched naked axes on which are borne terminal sporangia.

Rhyniophytoid. Any of several members of a possibly artificial group of early evolving (Early Silurian to Early Devonian) fossil plants having both bryophyte- and trachyophyte-like characters.

Ribonucleic acid (RNA). Usually single-stranded nucleic acid involved in protein synthesis in prokaryotes and eukaryotes, composed of four types of nitrogenous bases (adenine, guanine, uracil, and cytosine) bound to a linear sugar- (ribose-) phosphate "backbone."

Ribonucleotide. Nucleotide containing ribose sugar.

Ribose. Organic compound, a five-carbon sugar of the empirical formula $C_5H_{10}O_5$.

Ribosomal ribonucleic acid (rRNA). RNA that is a fundamental structural element of ribosomes.

Ribosome. Any of the RNA-rich cytoplasmic granules that are sites of protein synthesis.

Rubisco. Organic compound, ribulose bisphosphate carboxylase/oxygenase, the CO_2-fixing enzyme of photoautotrophs.

Saurischia. Zoological group of dinosaurs characterized by a reptilelike pelvis (see *Ornithischia*).

Sclerite. Physically resistant plate, piece, or spicule forming part of an exoskeleton.

Seed plants. Vascular seed-producing plants such as gymnosperms and angiosperms.

Sexual dimorphism. Occurrence in a single species of morphologically distinct males and females.

Sexual reproduction. Type of reproduction characteristic of most eukaryotes that involves meiosis (production of haploid spores or their equivalents) and syngamy (fusion of gametes).

Sheinwoodian Stage. Geological stage of the Middle Silurian.

Siegenian Stage. Geological stage of the Lower Devonian.

Silurian Period. Geological period of the Paleozoic Era (see Geological Timetable on the inner front cover).

Sipunculida. Zoological phylum that includes peanut worms.

Solar nebula. Gaseous and particulate cloud from which the sun and the planets of the solar system formed as a result of gravitational condensation.

Species. The fundamental category of biological classification, ranking below the genus and in some taxa comprised of subspecies, defined as "members of an actually or potentially interbreeding population, genetically isolated from all other such populations."

Sphaerocarpales. Botanical order of the liverworts.

Sphaeromorph. Any of numerous types of morphologically simple spheroidal acritarchs.

Sphagnaceae. Botanical family of the mosses.

Sporangium. Saclike structure, the site of meiosis (and therefore of spore production) in cryptogammic plants.

Spore. Haploid product of meiotic division in cryptogammic plants.

Sporophyte. In a plant life cycle, the diploid generation on which sporangia are borne (see *Alternation of generations* and *Gametophyte*).

Sporopollenin. Resistant organic matter forming the walls of spores and pollen.

Stage. Formally recognized stratigraphic subdivision of a geological period.

Stegosauria. Zoological group that includes plated ornithischian dinosaurs such as, for example, *Stegosaurus*.

Sterile axis. Plant axis devoid of fertile organs such as sporangia (see *Fertile axis*).

Sterome. Thick-walled peripheral outer cortical tissue providing physical support in some early evolving plants.

Stomate. Pore in a plant epidermis through which gaseous interchange takes place.

Strecker synthesis. Chemical reaction series via which amino acids are formed from aldehyde, hydrogen cyanide, and ammonium.

Stromatolite. Accretionary organosedimentary structure, commonly finely layered, megascopic, and calcareous, produced by the activities of mat-building communities of mucilage-secreting microorganisms, principally filamentous photoautotrophic prokaryotes.

Sugar. Any of various oligosaccharides, such as sucrose, fructose, and similar organic compounds, having the generalized formula CH_2O.

Supergroup. Major category in the stratigraphic classification of rocks ranking above the group.

Symbiosis. The living together in more or less intimate association or close union of two dissimilar organisms.

Sympatric. Pertaining to species living in the same area.

Synapsida. Zoological group that includes primitive (Pelycosauria) and advanced (Therapsida) "mammal-like" reptiles characterized by a single pair of temporal openings (fenestrae) in the skull (see *Anapsid* and *Diapsid*).

Syngamy. Fusion of gametes, as in eukaryotic sexual reproduction.

Tachytely. Unusually fast rate of morphological evolution (see *Bradytely*, *Horotely*, and *Hypobradytely*).

Taxon. In biological classification, a unit of any rank (i.e., a particular species, genus, family, class, order, or phylum or division) or the formal name applied to that unit.

Taxonomy. The scientific naming and classifying of organisms.

Telychian Stage. Geological stage of the Lower Silurian.

Template polymerization. Chemical reaction in which a macromolecule serves as a pattern for the formation of a polymer.

Terminal sporangia. Spore sacs borne on the upper ends of plant axes, as in the rhyniophytoids, rhyniophytes, and trimerophytes (see *Lateral sporangia*).

Tertiary Period. Earlier of two periods of the Cenozoic Era (see Geological Timetable on the inner front cover).

Tetrahedral tetrad. Pyramidal arrangement of four closely packed spheroids, as in the spore tetrads produced by many cryptogammic plants.

Tetrapod. Any of diverse terrestrial vertebrates having two pairs of limbs.

Thalloid. Consisting of or resembling a thallus, a plant body that lacks differentiation into distinct organs such as stem, leaves, and root.

Thecodontia. Zoological group of primitive archosaurian reptiles that probably included the ancestors of dinosaurs, crocodiles, and pterosaurs.

Therapsida. Zoological group of advanced synapsid "mammal-like" reptiles that includes the Anomodontia and Theriodontia (see *Synapsida*).

Theriodontia. Zoological group of carnivorous therapsid reptiles that includes the Gorgonopsia, Therocephalia, and Cynodontia (see *Anomodontia*).

Thermophilic. Of, or relating to an organism growing at a high temperature.

Therocephalia. One of three zoological subgroups of theriodont therapsid reptiles.

Theropoda. Zoological subgroup of obligately bipedal saurischian reptiles that includes all carnivorous dinosaurs such as, for example, *Tyrannosaurus* and *Compsognathus*.

Thymine. Organic compound, a type of pyrimidine and one of the five nitrogen-containing bases present in biological nucleic acids (see *Nucleic acid*).

Trace fossil. Type of metazoan fossil consisting of a track, trail, burrow, or similar structure.

Tracheid. Elongate tubular cell type characteristic of xylem and that functions in conduction and support.

Tracheophyte. Vascular tissue- (xylem- and phloem-) containing plants such as ferns, gymnosperms, and angiosperms.

Triassic Period. Earliest period of the Mesozoic Era (see Geological Timetable scale on the inner front cover).

Trilete mark. Y-shaped suture occurring on the proximal face of spores produced in tetrahedral tetrads.

Trimerophytina. Major botanical group (subdivision) that includes relatively early evolving (Devonian) naked or spiny monopodial or pseudomonopodial vascular plants on which are borne dichotomizing branchlets having trusses of terminal sporangia.

Ultrastructure. Subcellular structure discernable by electron microscopy.

Uracil. Organic compound, a type of pyrimidine and one of the five nitrogen-containing bases present in biological nucleic acids (see *Nucleic acid*).

Uraninite. Mineral, uranium oxide approximately of the composition UO_2.

Vascular tissue. Tubular conducting tissues, phloem and xylem, in vascular plants.

Vendian Period. Latest period of the Proterozoic Era (see Geological Timetable on the inner front cover).

Vertebrate. Chordate animal possessing a segmented backbone.

Vertebrate Bauplan. Generalized pattern, or "blueprint" of vertebrate anatomical organization.

Viviparous. Producing a fetus that is nourished within the maternal body (as in placental mammals), a specialized version of ovoviviparousness.

Warm-blooded. Informal term for endothermy.

Wenlock Stage. Geological stage of the Middle Silurian.

Xylem. Tubular, thick-walled, and commonly lignified water-conducting tissue of a vascular plant (see *Phloem* and *Vascular tissue*).

Zircon. Mineral, $ZrSiO_4$.

Zooxanthella. Any of various symbiotic, unicellular, micro-algal protists that live within the tissues of metazoans such as some corals and clams.

Zosterophyllophytina. Major botanical group (subdivision) of relatively early evolving (Devonian) vascular plants characterized by naked or spiny axes on which are borne lateral sporangia.

Zygote. The diploid cell formed by fusion (syngamy) of two haploid gametes, the earliest formed cell of the embryo of animals and of the sporophyte generation of plants.

INDEX